Private Doubt, Public Dilemma

Other Volumes in the Terry Lectures Series Available from Yale University Press

Private Doubt, Public Dilemma

Religion and Science
since Jefferson and Darwin

KEITH THOMSON

Yale

UNIVERSITY PRESS

New Haven and London

Published with assistance from the foundation established in memory of William McKean Brown.

Yale University Press books may be purchased in quantity for educational, business, or promotional use. For information, please e-mail sales.press@yale.edu (U.S. office) or sales@yaleup .co.uk (U.K. office).

Set in Janson Roman type by Tseng Information Systems.
Printed in the United States of America.

Library of Congress Cataloging-in-Publication Data
Thomson, Keith Stewart.
Private doubt, public dilemma : religion and science since Jefferson and Darwin / Keith Thomson.
pages cm. — (The Terry lectures series)
Includes bibliographical references and index.
ISBN 978-0-300-20367-7 (cloth : alk. paper) 1. Religion and science. 2. Darwin, Charles, 1809–1882. 3. Jefferson, Thomas, 1743–1826. I. Title.
BL240.3.T485 2015
201'.65—dc23

2014040042

A catalogue record for this book is available from the British Library.

This paper meets the requirements of ANSI/NISO Z39.48–1992 (Permanence of Paper).

10 9 8 7 6 5 4 3 2 1

The Dwight Harrington Terry Foundation Lectures
on Religion in the Light of Science and Philosophy

The deed of gift declares that "the object of this foundation is not the promotion of scientific investigation and discovery, but rather the assimilation and interpretation of
that which has been or shall be hereafter discovered, and its
application to human welfare, especially by the building of
the truths of science and philosophy into the structure of a
broadened and purified religion. The founder believes that
such a religion will greatly stimulate intelligent effort for the
improvement of human conditions and the advancement of
the race in strength and excellence of character. To this end
it is desired that a series of lectures be given by men eminent in their respective departments, on ethics, the history
of civilization and religion, biblical research, all sciences and
branches of knowledge which have an important bearing on
the subject, all the great laws of nature, especially of evolution . . . also such interpretations of literature and sociology
as are in accord with the spirit of this foundation, to the end
that the Christian spirit may be nurtured in the fullest light
of the world's knowledge and that mankind may be helped
to attain its highest possible welfare and happiness upon this
earth." The present work constitutes the latest volume published on this foundation.

Brethren, scholars, men of science, yours is a priestly calling.

ANDREW P. PEABODY, *The Connection between Science and Religion: An Oration Delivered before the Phi Beta Kappa Society of Harvard University*, August 28, 1845

Contents

Preface

Charles Darwin was never just a scientist; he was fascinated by metaphysics and tormented about religion. During the tense period in 1838 just before the final pieces of his theory of evolution by natural selection fell into place, just as he was about to propose to Emma Wedgwood, and just before his health failed miserably under the pressure of it all, he had been reading Wordsworth. Judging by entries that he made in his M Notebook, he had been thinking about the branch of theology called *natural theology*, according to which the beauty, order, and purposiveness of life on earth can be seen as proof of God's existence. Intrigued by his own aesthetic responses to nature, he made a set of telegraphic notes on the problem of assessing beauty, particularly the beauty of scenery, including this reference: "[See] Wordsworth about science being sufficiently habitual to become poetical."[1] There is a second reference to Wordsworth in a notebook from the end of 1856: "At end of Burke's essay on the sublime & beautiful there are some notes & likewise on Wordsworth's dissertation on poetry."[2]

In both cases, Darwin must have been referring to the *Preface to the Lyrical Ballads* (1800), where Wordsworth wrote:

> The Man of science seeks truth as a remote and unknown benefactor; he cherishes and loves it in his solitudes: the Poet, singing a song in which all human beings join with him, rejoices in the presence of truth as our visible friend and hourly companion. . . . It is the impassioned expression which is the countenance of all Science. . . . If the labours of men of Science should ever create any material revolution, direct or indirect, in our condition, and in the impressions which we habitually receive, the Poet will sleep then no more than at present, but he will be ready to follow the steps of the man of Science, not only in those general indirect effects, but he will be at his side, carrying sensation into the midst of the objects of the Science itself. . . . If the time should ever come when what is now called Science, thus familiarized to men, shall be ready to put on, as it were, a form of flesh and blood, the Poet will lend his divine spirit to aid the transfiguration, and will welcome the Being thus produced, as a dear and genuine inmate of the household of man, and science will become part of an "atmosphere of sensation" in which we all "move our wings" and create a material revolution, direct or indirect, in our condition.[3]

In singling out this passage from the *Preface* Darwin asked a central question, relevant both then and now. What role does science play in our lives? Two centuries after Wordsworth, it seems that we have come a long way in the familiarization and habituation of science toward making that "material revolution in our condition." That science has become poetical might well be challenged, but there can be no doubt that it has completely and irrevocably changed the whole atmosphere "in which we all move our wings"—and not always in ways that are comfortable and secure. And this is the situation that Dwight Terry has asked us to ponder in his series of lectures on religion and science, to which this book is a contribution.

Acknowledgments

I am extremely grateful to Yale University, the Dwight Harrington Terry Foundation, and the Terry Lectures Committee for the invitation to present the Terry Lectures for 2012. Lauralee Fields, associate secretary at Yale University, Jean Thomson Black (no relation) at Yale University Press, and Professor Dale Martin were most gracious hosts, infinitely patient in adjusting the eventual scheduling to suit my personal situation. My only regret is that my friend and colleague Professor Leo Hickey (now deceased) was not well enough to attend the presentations that he did so much to encourage.

These lectures, now elaborated into essays, in one sense represent a distillation of ideas that I have been harboring and refining for much of my academic life as a biologist and historian of science. The result is that the number of people whom I should acknowledge for their inspiration, advice, and criticism is almost endless, and there are far too many to list here. But they know who they are and the extent of my gratitude. And in that list I include all the librarians I have known, especially those at Harvard, Yale, the Academy of Sciences of Philadelphia, the University of Oxford, and the American Philosophical Society (among many, many others). In this digital age, a world without libraries is for me unthinkable. The work of students of every stripe is, however, made infinitely easier by the scholars and publishers who have made freely available in digital form the edited papers of Jefferson (Princeton University Press), Darwin (Cambridge University Press), and many others.

I have to single out Andrew O'Shaughnessy (director of the International Center for Jefferson Studies at Monticello) for having encouraged me into ever-broader studies of the enigmatic Thomas Jefferson, Professor Stan Rachootin (Mt. Holyoke College) for having pushed me, some thirty-five years ago, metaphorically into the arms of the equally enigmatic Charles Darwin, Professor Robert Ryan (Rutgers University) for guiding me through Wordsworth's *Preface*, and my late father the Reverend Ronald William Thomson for having instilled in me a love of the King James Bible.

My wife, Linda Price Thomson, and our daughters, Jessica and Elizabeth, have always been an inspiration, while Olivia, Jane, and William are perfectly wonderful grandchildren. I thank Linda and my ever-patient assistant for tirelessly combing the manuscript for errors; they could not possibly have caught them all.

CHAPTER ONE

The Long-Standing Problem

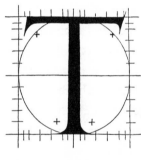HE wording of the Terry Lectures offers a daunting challenge: to illuminate "the assimilation and interpretation of that which has been or shall be hereafter discovered, and its application to human welfare, especially by the building of the truths of science and philosophy into the structure of a broadened and purified religion. . . . Such a religion will greatly stimulate intelligent effort for the improvement of human conditions and the advancement of the race in strength and excellence of character."

One cannot read these words without recognizing the extent to which difficulties in the relations between science and religion, rather than their useful cooperation or at least coexistence, have taken over discussion of the two subjects. Today the one thing that everyone seems to know about science and religion is that they often intersect antithetically, even antagonistically, but rarely positively. At best there may exist a sort of mutual tolerance. An enormous amount has been written on this subject, from every point of view. Between action and reaction it is hard to decide where any particular argument starts or ends, and frankly it is all becoming more than a little tiresome, with many commentators on the subject making things worse rather than better.

With that in mind, I would like to start by recalling some commonsensical words written in 1860 as the preface to a series of essays published as *Essays and Reviews*. The authors of this highly controversial book, possibly as consequential in its time as Charles Darwin's *On the Origin of Species*, were seven senior prominent clerics and academics who discussed the impact that recent developments in science and biblical study were having in changing the face of common belief. Variously, they summarized new research challenging the literal truth of Genesis (and therefore the revelatory truth of the Bible), they canvassed for accepting an ancient age for the earth, and one author even endorsed evolutionary theory in the form of Darwin's natural selection.

"The Volume, it is hoped, will be received as an attempt to illustrate the advantage derivable to the cause of religious and moral truth, from a free handling, in a becoming spirit, of subjects peculiarly liable to suffer by the repetition of conventional language, and from traditional methods of treatment."[1] I share with the anonymous author of this slightly irritable statement a feeling that we are suffering today, just as they were in 1860, from a repetition of conventional phrases, thinking, and argumentative positions, and I do not imagine that I can add anything to the philosophical debates that have themselves produced much noise but little light over the last two millennia.[2] In what follows, however, it may be that a "free handling" of some of the issues and a "becoming spirit" may be useful even if it means, as it did in 1860, risking offending some.

I will attempt to avoid the sirens of both antireligious and antiscience rhetoric and stress something closer to Mr. Terry's initial goal. At least I shall try to avoid the behaviors that have become so familiar to us recently—the shrill stakeholding of some participants on both sides, and the revisionism of some historiographers and the apologetics of others. But inevitably the conflict cannot be avoided totally. As a scientist, I am not qualified to pronounce on theology or religion, but as a historian of science, I can try to trace some events, characters, and debates of the past and to see how they impinge on our lives in the present highly troubled times. While science and religion, however construed, seem to have gone their separate ways for so long as almost no longer to recognize each other, I will try to show that there is just as pressing a case today as there was in the eighteenth and nineteenth centuries for members from the two sides to find their way toward common ground.

I cannot avoid mentioning here that the "disconnect" between science and religion is not unique. There is and always has been, as Wordsworth's *Preface* shows so clearly, a tension between the artistic and the scientific viewpoints, and indeed also a tension between the artistic and the theological. Nor are the difficulties between religion and science solely a product of the nineteenth and twentieth centuries. Each age has had its own crisis—in the sixteenth and seventeenth centuries there had been perhaps the most dramatic scientific discovery of all: a convincing demonstration that the earth was not after all the center of the universe, leading humans to the possibility that our role on the cosmic stage is vanishingly small and insignificant. In the eighteenth century the biggest difficulties were focused on scientific approaches to Creation and the age of the earth. As Benjamin Rush noted in his Commonplace Book around 1810, "Geology and botany have spoiled revelation."[3]

We biologists can become testy when charged with having made Darwinism the locus classicus of the religion-science debate. It is important, we insist, to remember that the philosophical differences between material and immaterial approaches to the problem of existence form probably the oldest schism in philosophy, traceable to the pre-Christian Classical Age. Nonetheless, there was a great deal of truth in what an anonymous reviewer of Darwin's *The Descent of Man* wrote in 1871:

> We may . . . say what will be admitted on all hands, that the question raised by Mr. Darwin as to the origin of species marks the precise point at which the theological and scientific modes of thought come into contact. . . . [On] the mode in which divines and philosophers will ultimately reconcile their differences depends in great measure the future of human thought. Religion undoubtedly corresponds to an ineradicable instinct; and we can have no fear that religion itself will permanently suffer from scientific discoveries; it is quite possible, however, that the current religious ideas may be materially modified in conception of the current external world changes, and it is therefore well worth while to give some attention to this debatable land in which so many vigorous blows are being exchanged by the contending parties, previous to the final reconciliation which we may confidently anticipate.[4]

We live in especially troubled times for the relationships between the sciences and religions (note the use of the plurals). The competition for authority has rarely if ever been so overt and political. And it is a curious and frustrating thing that science, which above all is a search for truth and hard uncomfortable facts and fixed laws, is the biggest source of change that constantly challenges convention. I can give no better example of changing science than the discovery in the last decade or so that the cosmos, as we thought we knew it, actually consists mostly of dark matter or dark energy. And the universe, which had been thought to be constant in size, was next found to be expanding at a constant rate, except that the contemporary view is that it is in fact accelerating. Meanwhile, on earth a torrent of developments in biomedical science continues to upset theological applecarts.

In part, my aim here will be to show that, nonetheless, our modern problems are not so different in kind from those experienced by our recent forebears. We set off shakily into an Age of Reason where, as now, reason was just as often confounded both by the prejudice of what we thought we already knew and by plain ignorance. Then came the Age of Science — of discovery, description, and explanation; it was reason put to work. The Age of Science did not replace the Age of Reason but was simply (or not so simply) layered onto it — it was the figurative promise of the atomists exemplified and eventually made literal truth. In more modern times, yet a third age, the Age of Uncertainty and Doubt, has been layered onto reason and science. Its patron saints are Heisenberg and Foucault; its agents are intellectual honesty, clouded by a dangerous mixture of politics and religion, and the pervasive processes of change in everything we know.[5] The result, as my old teacher at Harvard Alfred Sherwood Romer once said, is that everywhere "with increasing knowledge comes triumphant loss of clarity."

An earlier Terry Lectures volume asked why the debate between religion and science continues.[6] I want to look at things from a different viewpoint. These lectures will have a number of broad subthemes. First, I will emphasize that the science-religion predicament is but a small part of a much broader set of difficulties and opportunities that arise wherever and whenever new knowledge is added to old. I will follow up on an idea presented by the first Terry lecturer, J. A. Thomson, that there is a cost (he said a tax) associated with new knowledge.[7] There may also be huge dividends, but adding new knowledge to our personal hard drives is not always easy; there are costs involved in the form of old ideas shed and new loci of disagreement discovered. Also, some of what seem to be modern ideas and positions are not new at all but have long genealogies.

A key problem for any age is dealing with change. This, I believe, is at the root of much of our present difficulty, and the main point I want to dwell on here is that the celebrated conflict between religion and science is really part of a much broader phenomenon occurring whenever there is change in our knowledge—either or both in what we know and the context in which we know it. There is not always outright conflict, but there is always a dilemma—what shall we do when confronted with new knowledge? How do we adjust to a new style of modern painting or music, a geological discovery, or even a new translation of the Bible?

Sometimes, and for some people, a challenging new idea is quickly assimilated. When Thomas Henry Huxley first read Darwin's *On the Origin of Species*, he complained, "How stupid of me not to have thought of that." Needless to say, it was not the response of everyone. "Over my dead body" may still be the more common response. Sometimes our response is contorted efforts to "save the phenomenon"—to have it both ways. One favorite example is that of Ptolemy and the epicycles he had to invent to explain why planets sometimes seem to go backward. Perhaps the commonest approach is to do nothing and wait to see what happens. And the irony is that we want our conventions and authorities to reflect the eternal verities and at the same time also to be responsive to the ever-changing present.

One of the reasons why, then and now, there has been so much room for disagreement over the role of science and its relationship to religion is that both are at best works in progress. Epistemologists and philosophers tell us that, despite our best efforts, what we know is a curious mixture of what we know by personal experience and what we accept from the work of others. And, here's the rub. What we know for ourselves, or information that we are presented with by others, always changes much, much faster that does what we can call conventional thinking and institutional authority.

At any moment what we know—what we think we have validated for ourselves and what we accept as true—is in a state of flux. Our knowledge is a mixture of ancient and modern.

We are each a complex mixture of influences from the latest discoveries and creations to the classics, the Bible, folklore, and the influences of, say, Middle Eastern mathematics and Asian contemplation. We have a legacy in Roman law and owe an enormous debt to the Greeks, who touched "with light of reason and grace of beauty the wild Northern savages."[8] The wooly-seeming term *happiness*, for example, the pursuit of which was enshrined by Jefferson in the American Declaration of Independence, was perfectly defined by Aristotle in his *Nichomachean Ethics* (circa 350 BC).[9]

While we struggle to shed the old, we are almost weighed down by what is brand-new. And we are the creation of our institutions and authorities—simply because we cannot parse out for ourselves every philosophical position, authenticate every mathematical proof, or understand quantum mechanics. We "accept" an enormous amount of what we "know." But we *want* to understand; we need, wherever possible or practical, to authenticate things for ourselves.[10] The crux of any conflict—or indeed any convergence—between the sciences and religions is the challenge of the new and the inherent conservatism of the old. On some scale we all experience it.

We often think of religion (and theology) and science as two great behemoths of theory and practice, inevitably butting heads. However, I think one key to understanding all this, and potentially dealing with it, lies not only in studying the institutions, powerful as they are, but also, and first, examining what happens when *individuals* are confronted with new knowledge that does not fit comfortably with the old.

When intellectual conflict arises, it happens first in the minds of individuals, rather than in societies or institutions.

The exciting moment is not *just* Galileo facing the tribunal—it is the young Galileo first reading Copernicus and Kepler and realizing the consequences. Or it is the young Copernicus, having already toyed with the idea that the sun was the center, not the earth, going to Krakow to study with Adalbert Brudzewski, the great expert on Ptolemy. As the author John Banville imagined it, "There had been distilled one tiny precious drop of pearly doubt. He could no longer remember where or when he had found the flaw, along what starry trajectory, on which rung of those steadily ascending ladders of tabular calculation, but once detected it had brought the entire edifice of a life's work crashing down with slow dreamlike inevitability. Professor Brudzewski knew that Ptolemy was gravely wrong."[11]

I must begin with a bold statement and then qualify it. First, science and religion are deeply in conflict over the role of any supernatural being in the conduct of the affairs and processes, properties and laws of the material world. Second, various accommodations and connections are possible. The work of too many historians, and particularly historiographers, however, in writing about science and religion has seriously conflated and confused two aspects of the situation—the philosophical and practical. In these essays I want to look at how real people dealt with the issues of religion and science in their lives. And then, perhaps, having straightened things out a bit, we can return to the "conflict" issue and discover what the two sides have in common and how they can actually work together.

I will look at the issue in narratives of the lives of two great men: Thomas Jefferson and Charles Darwin, two men whose intellect, work, and lives I mostly admire (but not without certain reservations, as will probably become apparent). Benjamin Disraeli will make the occasional appearance.

I began my academic career as a biologist but have become more of an historian. Part of my historical work has been to try to place Darwinian science into its larger context through relating Darwinism to the ideas about evolution that preceded Darwin, and often in a geological rather than biological context. I have also been fascinated by Jefferson's reaction to, and contributions to, the science of his age. I am constantly struck by the fact that his greatest contributions to science, written out in *Notes on the State of Virginia*, were produced a mere fifty years before Darwin's *Beagle* voyage, and yet their two worlds were different in every possible way.

After laying some groundwork on religion and science here and in the following chapter, I will take the twin examples of Jefferson struggling to reconcile his science with his religion and Darwin trying to accommodate his religion to his science to demonstrate some broader issues of how matters of individual conscience intersect with authority, especially religious authority. And how difficult it is for any of us to change. I must make it clear, however, that my references to religion are focused on the Judeo-Christian tradition in which I grew up, and I claim no expertise in subjects such as Islam— an omission that I must leave to others to repair.

In Thomas Jefferson's lifetime, science was on the cusp of the changes that propelled us into modern society, both in understanding of the world we live in and the ways that it could be managed. In my first example, we will see Jefferson struggling with questions he could not answer—he always hated that. It would, however, be unreasonable for us to expect him to have solved the problem that baffled him—the reconciliation of modern geological science with the traditional account of Creation in the first chapter of Genesis.

Charles Darwin lived during an even greater flood of scientific discovery. His problem, like Jefferson's, was intensely personal but had global ramifications. He had discovered a whole new scientific principle and system that ran counter to conventional religious belief and orthodox scientific thinking. He accommodated this conflict by compromising, and then bitterly regretted it. So he paid at least two prices—the price of abandoning his faith and setting himself up against a whole culture, and then the price of not sticking to his convictions.

Both men were berated for being too independent thinkers on science and religion and for rejecting miracles and all the trappings of Christianity.

In the famous Wilberforce-Huxley debate at Oxford in 1860 and the lesser-known debates in Boston and Cambridge, Massachusetts, some time before, we see how the issue of science and religion becomes reified as different groups strive for new authority or to preserve the old.

All this will lead us to a discussion of how this contest among rival authorities affects us today. F. Scott Fitzgerald in *The Crack-up* (1936) famously wrote, "The test of a first-rate intelligence is the ability to hold two opposed ideas in the mind at the same time, and still retain the ability to function." But in fact most of us (and of course the readers of these essays) are nothing like Buridan's ass that metaphorically starved between a bale of hay and a pail of water. Many of us can cope with paradox, uncertainty, and confusion and are much more like—well—Thomas Jefferson. But first I must digress.

CHAPTER TWO

Religion and Science

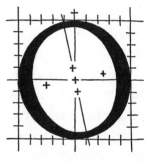NE of the most marvelous things about science, whether for the direct participant or the consumer, is its constant capacity to surprise and delight us, to produce truly epochal change and also to illuminate the smallest corner of existence. And in the process of learning so much more about the world and how it works, including the function of our own bodies, I find it fascinating that we have not lost a sense of wonder. Staring at the night sky—in places where it is not obscured by light pollution—although we know so much about what is out there, we really know very little, and what we do know is very curious. Is there anyone who has lost his or her amazement that the disk of the moon fits so exactly that of the sun in a solar eclipse?[1]

Of all the great theoretical innovations in science, the one that I admire most and wonder at most often is the work of the

13

pre-Socratic Greek atomists Leucippus and Democritos in the fourth and fifth centuries BC. Their insights into the very nature of matter are encapsulated in two familiar quotations from Democritos: "Nothing exists except atoms and empty space; everything else is opinion" and "Everything existing in the Universe is the fruit of chance and necessity" (that is, of chance and law). It is not condescending or patronizing and certainly not triumphalist (all qualities historians of science are constantly charged by critics to possess) to observe how remarkably prescient the atomists were, even though they had no notion of why atoms moved or what they were. Naturally enough, such a philosophy was dangerous; atoms, chance, and necessity neither explained nor required free will. In the early nineteenth century, the French mathematician Laplace put this clearly in his *Philosophical Essay on Probabilities* (1825): "We ought to consider the present state of this universe as the effects of its previous state and as the cause of that which is to follow. An intelligence that, at a given instant, could comprehend all the forces by which nature is animated and the respective situation of the beings that make it up, if moreover it were vast enough to submit these data to analysis, would encompass in the same formula the movements of the greatest bodies of the universe and those of the lightest atoms. For such an intelligence nothing would be uncertain, and the future, like the past, would be open to its eyes."

The strict atomist view left no room for anything outside the material world. The most popular alternative view, obviously, is that the world is something far greater than the sum of its measurable parts. It has been created by and even is daily controlled by supranatural powers, gods, and is an expression of their will, their design, and their purpose for us. The most marvelous thing about any religion is not just its capacity to inspire in the way that science can, but the way it can also lead us to worlds outside of ourselves, to articulate modes of moral behavior and to provide a discipline for their enaction.

How one might find an independent proof of the existence of any god has always been an insoluble question, both for science and religion. Experiments on the power of *supplicative* prayer have a long history, for example.[2] They necessarily consist in attempts (prayers for the sick or for rain) to trap God into revealing himself, which is something that is both blasphemous and hopeless, if only because it is possible that God, as has been said, hears our prayers and always answers, "No." *Meditative* prayer is often rewarding, but whether it is God who provides the rewards is similarly impossible to determine.

Failing some kind of proof, the evidence for the reality of God (or gods) has most often been taken to be the extraordinary complexity and diversity of the natural world, to which various writers have added its beauty, its marriage of form and function, and its utility for human life. And yes, that coincidence of the sizes of the moon and sun viewed from the earth. Such a perfect world must have been specially created, and that required the prior existence of an intelligent creative being—God. The titles of books by William Paley (*Natural Theology; or, Evidences and Attributes of the Deity* [1801]) and John Ray (*The Wisdom of God Manifest in the Works of His Creation* [1691]) sum it up. The basic idea of natural theology was not new in the seventeenth century, however; it traces back to Aristotle and was familiar two thousand years ago. Epictetus put the issue succinctly in Book One of his *Discourses:* "Any one thing in the creation is sufficient to demonstrate a Providence to a humble and grateful mind."

The Roman orator and philosopher Marcus Tullius Cicero perhaps expressed it best in his *De Natura Deorum*, written as an inquiry, a dialogue among Stoic, Epicurean, and Skeptic philosophers about, as the title says, the nature of the gods. The Stoic Balbus asked,

> How then can it be consistent to suppose that the world, which includes both the works of art in question, the craftsmen who made them, and everything else besides, can be devoid of purpose and of reason? . . . [If] the products of nature are better than those of art, and if art produces nothing without reason, nature too cannot be deemed to be without reason. When you see a statue or a painting, you recognize the exercise of art; when you observe from a distance the course of a ship, you do not hesitate to assume that its motion is guided by reason and by art; when you look at a sun-dial or a water-clock, you infer that it tells the time by art and not by chance; how then can it be consistent to suppose that the world, which includes both the works of art in question, the craftsmen who made them, and everything else besides, can be devoid of purpose and of reason?
>
> Suppose a traveller to carry into Scythia or Britain [note that Cicero wanted to use as an example the most primitive society he knew of] the orrery recently constructed by our friend Posidonius, which at each revolution reproduces the same motions of the sun, the moon and the five planets that take place in the heavens every twenty-four hours, would any single native doubt that this orrery was the work of a rational being?[3]

An even better example of how old are modern antievolutionists' arguments comes from something else Cicero wrote on the same subject. A familiar modern charge against the role of chance in evolution is that if one were to give a monkey a typewriter, no matter how long it labored, it would never produce even one intelligible phrase, let alone a line from Shakespeare. Cicero, arguing against the atomists, said the same thing: "Must I not marvel that there should be anyone who can persuade himself that there are certain solid and indivisible particles of matter borne along by the force of gravity, and that the fortuitous collision of these particles produces this elaborate and beautiful world? I cannot understand why he who considers it possible for this to have occurred should not also think that, if a countless number of copies of the one-and-twenty letters of the alphabet, made of gold or what you will, were thrown together in some receptacle and then shaken out on to the ground, it would be possible that they should produce the Annales of Ennius. I doubt whether they could possibly succeed in producing a single verse."[4]

Instead of an orrery, William Paley built his natural theology around someone finding a pocket watch lying on the ground. Anyone, even if that person did not know what it was for, would know that it must have had a designer and maker. If that is true for a watch, it must surely be true, he argued, for something as complicated as a living organism. For Paley (and all his modern followers), natural theology gathered up the weight of scientific observation and scientific understanding of the material world to demonstrate the power and goodness of the Creator. Furthermore, just as the watch has a purpose, so living creatures are adapted to carry out particular functions—the webbed feet of water birds, the beak of a hummingbird—and this demonstrates one of the key elements in a religious belief, namely, purpose (teleology). Today the "theology" of *intelligent design* depends on the same old complexity argument, denying that complex molecular structures and mechanisms could have self-assembled through a history, however intricate, of chance events among simpler molecules. Natural theology famously failed to answer the question, however, of why such a benevolent and careful God created or allows the whole panoply of evil and disease with which his Creation is so prominently beset.

A fancy variant of all this exists in the various forms of an *anthropic principle*. We know that the material composition and laws of the cosmos are such that, at least on one tiny speck of it, life emerged. As has been pointed out, this leads to a tautology: "This universe which contains ourselves must be compatible with our having appeared in its history."[5] It is easy to proceed from this to the inference that somehow the appearance of life—just as we know it—was inevitable (and so, presumably, was that wonderful convergence of the relative size of the sun, moon, and earth and their orbits). And if any of that were true, the next leap would be that somehow "there is a divine purpose behind this fruitful universe."[6]

The problem is, of course, that we know only what our own sense systems allow us to perceive and our own experience interprets. We can scarcely be surprised, therefore, that the universe seems to us to be so compatible with our observations. We also cannot know whether we might simply be a dreadfully failed four and a half billion year "experiment" that merely produced us, while the really interesting results are off in some other corner of the universe. Similarly, on earth the slightest tweak of parameters over cosmological history might have, over time, produced a different result. Perhaps the most serious flaw of "anthropic" thinking is that it proceeds from the premise that evolution is aimed at us and stops with us: that we humans are the target, purpose, and endpoint of history; we are God's purpose. This human-centric view of evolution is rendered untenable by the fact of evolution which, among other things, explains why we, the supposed pinnacle of Creation, are physically so imperfect.[7] Still in the process of evolving from apes, our human females have a birth canal too small for the huge crania of our fetuses; our shortened jawlines have no room for the last teeth to erupt, and so we have impacted wisdom teeth; in imperfectly taking on upright posture, we have a weak weight-bearing construction of the sacroiliac.

All of which gets us precisely nowhere; there might be a God somewhere with a special purpose, or playing dice, or there might not. The anthropic principle, as Darwin said of Special Creation, seems only to be "restating the [problem] in dignified language." It does not advance us from the most elemental situation: belief in a supreme being remains a belief. What you do with it is a very personal decision. And we must remember also that some philosophies of both science and religion leave open the possibility that it is all an illusion.

There is a danger in proceeding further without having defined some terms, particularly *science* and *religion*. The approach of many recent works on religion and science has been to deny that there is any essential conflict between the two, merely a series of local flare-ups. This approach depends in great part on the entirely reasonable assessment that science and religion are such immensely complex, context-dependent entities that attempts to define some *essence* of either one are counterproductive and probably obfuscatory. Such works therefore studiously avoid defining the very subjects they discuss. I, on the other hand, have the old-fashioned notion that you, the reader, ought to know what I am talking about.[8]

At the risk of huge political incorrectness, let me pose some definitions, if not didactically, at least heuristically. With use of these sorts of definition, or with the definitions themselves, at least there will be no ambiguity as to what I refer. In such cases I usually rely upon my old friend the first edition of the *Oxford English Dictionary*, which reassuringly tells us that religion starts with the belief that there is a God who controls our lives and destiny and who is worthy of worship. Fundamental mental and moral attitudes result from this belief, as well as standards for a spiritual and practical life. In all religions there is an element of particular rites and observances, and many symbols. There is still merit in the very old distinction of three aspects of conventional religion. There is the *intellectual view* that religion is a way of knowing, regardless of particular creeds. In the *moralist view*, religion is a practical (or even tactical) matter, concerned with the direction of the will. The *romantic view* involves the reaction to—a "feeling" of—the presence of the Creator.

We should remind ourselves also of the three kinds of traditional arguments for the existence of God: *ontological*, starting with the idea of God in the mind; *cosmological*, the cosmos shows there must be a first cause; and *teleological*, purposiveness in the world shows purposive mind. All these assume a God. But there are other definitions of religion, and for some God is unnecessary. For example, Clifford Geertz defined religion as a system of "symbols, powerful and long-lasting motivations, and conceptions of a general order of existence clothed in an aura of factuality."[9] The issues with which these essays will deal, however, involve what some call merely a "popular assertion"—that there really is a God who is worthy of worship.

This is perhaps all old-fashioned stuff, as is my definition of science, where again I rely on the old *Oxford English Dictionary*, which defines science in terms of a *method of inquiry* applied to *organized knowledge*, leading to the discovery of *general laws* and restricted to those branches of study that relate to the phenomena of the *material universe* and their laws. As Condorcet put it in *The Future Progress of the Human Mind* (1796), "The sole foundation for belief in the natural sciences is this idea, that the general laws directing the phenomena of the universe, known or unknown, are necessary and constant."

Not all modern science is as reductionist as the atomist approach would suggest. There is plenty of room for emergent phenomena and processes. But most scientists today are methodological naturalists who do not necessarily deny the existence of God but operate under the assumptions that nothing they study in the material world is susceptible to influence by supernatural phenomena like miracles and that their studies cannot be shaped by the hand of God. By contrast, metaphysical naturalists deny the very existence of any kind of God.

As I have put it elsewhere, "Science does not deal with situations where observations cannot be made directly, measurements taken, or experiments performed. It therefore resists or rejects the world of miracles and the supernatural, demanding of it the same proofs as apply in the material world. If, one day, the sun actually were to be observed to move ten degrees backwards (II Kings 20, VII), Copernicus, Galileo, and every cosmologist after them would have been wrong."[10] And, it must be added, since the earth's rotation would have stopped, the results would have been disconcerting, to say the least.

David Hume, in *An Enquiry concerning Human Understanding*, addressed things perfectly: "If we take in our hand any volume of divinity or school metaphysics, for instance, let us ask, Does it contain any abstract reasoning concerning quantity or number? *No.* Does it contain any experimental reasoning concerning matter of fact and existence? *No.* Commit it then to the flames, for it can contain nothing but sophistry and illusion."[11]

I was reminded of this when I read a letter to the *New York Times* from the chief communications officer of the so-called Creation Museum in Kentucky justifying its approach to scientific data and, among other things, its presentation of exhibits showing dinosaurs and humans living together and claiming that the world is only six thousand years old. "Accepting the Bible as God's literal truth doesn't mean that we discount science," he wrote. "It does mean that we interpret scientific evidence from the biblical viewpoint. . . . Evidence isn't labeled with dates and facts; we arrive at conclusions about the unobservable past based on our pre-existing beliefs."[12] This statement, I think, fairly qualifies in Hume's terms as sophistry and illusion.

Finally, I must declare here my scientist's philistinism. Debates about whether there really can be objective knowledge are all very well, but in order to get anything done, a scientist must assume that there is. Evidence without dates and facts is mere opinion. As Emily Dickinson put it, "Faith is a fine invention / For gentlemen who see / But microscopes are prudent / In an emergency."

In the same pragmatic spirit, I will state here and now that arguments about whether there is a God or not are irrelevant in one sense—if enough people believe there is a God and act on their beliefs, then that becomes a political fact that has to be accommodated. It doesn't mean that the efforts of some to convert the religious from their worst excesses are wrong, and it doesn't mean that the efforts of others to convert the unbelievers to religion (usually some particular shape or shade of religion) are right. For our present purposes—this not being a philosophical exercise—let us just see what we can do with what we have got: societies divided both among and within themselves, and note that above all what we need is more and better education in both science and religion.

CHAPTER THREE

Mr. Jefferson's Dilemma

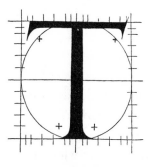HOMAS Jefferson's intellectual heroes were Bacon, Newton, and Locke. Mine are Jefferson and Darwin. Both men lived in rapidly changing times for every aspect of society, including science and religion. Much of what we understand about the two men is paradoxical and frustrating to anyone who likes to spell out a simple morality play in the lives of their heroes. However, if they lived in troubling times, then so do we; and perhaps we can learn from them.

Jefferson was the most complex of men, a deeper thinker than any president except perhaps Madison, an intellectual hero (but not a military one) of the War of Independence, and by turns charming, passionate, ruthless, admirable, and despicable. Throughout his life Jefferson had a keen interest in science, which in his day was a combination of natural philosophy and natural history. He can fairly claim to have been at least a co-founding father of American climatology and geography, scientific archaeology, and paleontology. His published contributions to science are principally contained in *Notes on the State of Virginia* (published in final form in 1787), which he wrote between 1782 and 1783.[1]

Notes was not just a compendium of Virginia geography, laws, and customs; it was also a record of the known natural history of America. Most of all, it was Jefferson's personal manifesto and thus tells us a great deal about him. Jefferson's stay in Paris as American minister between 1786 and 1789 gave him access to the cream of European scientific thinking, and the results of that show up in the flood of original ideas and observations contained in his voluminous later correspondence.

It is fascinating that the man whom we so revere for his writing of the Declaration of Independence could have been so vilified in the election of 1800. In 1782, the Marquis de Chastellux visited Monticello and found "a man, not yet forty, tall, and with a mild and pleasing countenance, but whose mind and attainments could serve in lieu of all outward graces; an American, who, without ever having quitted his own country, is Musician, Draftsman, Surveyor, Astronomer, Natural Philosopher, Jurist, and Statesman . . . Philosopher . . . a gentle and amiable man."[2] By contrast, after his presidency Jefferson was slandered: "To make the best of him, [he] was nothing but a mean-spirited, low-lived fellow, the son of a half-breed Indian squaw, sired by a Virginia mulatto father, as was well known in the neighborhood where he was raised, wholly on hoe-cake, bacon, and hominy, with an occasional change of fricasseed bullfrog, for which abominable reptiles he had acquired a taste during his residence among the French."[3]

As with everything else in his life, science could be triumph or failure for Jefferson. And often these peaks and valleys corresponded with popular tolerance or denial of religious philosophies like his. In all his intellectual life Jefferson searched for basic facts, truths and laws upon which to build a philosophy, and he was attracted therefore to Newton above anyone else in the world of science. He first read Newton as a student under William Small at the College of William and Mary, and his library, fully one-third of which consisted of books on science, shows his continued devotion to the work of the great master. As we will quickly be reminded, however, there are few truths that do not intersect with other truths, and often we find them to be contradictory. Jefferson's philosophical world was full of such contradictions; one such existed with respect to his ideas on geology.[4]

A deist, Jefferson was not a churchgoer, but religion was important to him philosophically. He believed firmly in the truth of the Old Testament but held that Jesus was a great teacher and healer, not divine. Like so many in the Age of Reason, he totally rejected the concept of miracles. He famously created a redacted version of the New Testament Gospels, with all the supernatural elements eliminated. And he held firmly to the basic premise of natural theology that the world, and especially life on earth, is so complex that it could not have arisen by any combination of "chance and necessity" but required a divine Creator. Toward the end of his life, he wrote a simple summary of his natural theology for John Adams:

> I hold (without appeal to revelation) that when we take a view of the Universe, in it's parts general or particular, it is impossible for the human mind not to percieve and feel a conviction of design, consummate skill, and indefinite power in every atom of it's composition. The movements of the heavenly bodies, so exactly held in their course by the balance of centrifugal and centripetal forces, the structure of our earth itself, with it's distribution of lands, waters and atmosphere, animal and vegetable bodies, examined in all their minutest particles, insects mere atoms of life, yet as perfectly organised as man or mammoth, the mineral substances, their generation and uses, it is impossible, I say, for the human mind not to believe that there is, in all this, design, cause and effect, up to an ultimate cause, a fabricator of all things from matter and motion, their preserver and regulator while permitted to exist in their present forms, and their regenerator into new and other forms.[5]

And Jefferson's language naturally reminds us of Newton's words on the same subject: "This most elegant system of the sun, planets, and comets could not have arisen without the design and dominion of an intelligent and powerful being."[6]

While traditional Christians believed that the path to enlightenment was to accept God's revelation and interpret nature in that light, Jefferson believed the opposite. The study of nature was the key, the road, to discovering the pure truths about God. And he saw the natural world of America as a particular gift of God and therefore a privileged opportunity.

In this essay I want to concentrate on Jefferson's ideas about geology and how he tried to relate them to his version of natural theology. There can be no doubt that in this, as in a lot of his scientific thinking, he was motivated by a negation, specifically a reaction against the ideas of the great French scholar the Comte de Buffon. His greatest contribution to natural science was his point-by-point rebuttal in *Notes on the State of Virginia* of Buffon's notorious opinion that American nature and native peoples were inferior and that the land and climate (both north and south) were unsupportive of anything but a primitive existence in which immigrant European stocks (people and livestock) degenerated.[7]

Buffon was more prescient (and better informed) as a geologist/cosmologist than as a naturalist of the Americas. With others, he had conceived of the earth as having cooled from an original molten ball of iron, probably spun off from the sun. He performed experiments on the rate of cooling of white-hot cannon balls to calculate that the earth must be at least seventy-five thousand years old. As it rotated and cooled, the molten mass necessarily assumed its characteristic oblate spheroidal shape, and that shape seemed to be firm evidence of a physical process at work.

In a letter to his friend the polymath Charles Thomson, secretary to the Continental Congress, Jefferson rejected all this in favor of the biblical account of Creation. "I give one answer to all these theorists. That is as follows: they all suppose the earth a created existence. They must suppose a creator then; and that he possessed power and wisdom to a great degree." Arguing very much like a lawyer with respect to the oblate spheroid shape of the earth, he turned Buffon's argument upside down. If an oblate spheroid was the correct shape for a revolving mass, then the Creator would certainly have

made it that way. "I suppose that the same equilibrium between gravity and centrifugal force which would determine a
fluid mass into the form of an oblate spheroid, would determine the wise creator of that mass, if he made it in a solid
state, to give it the same spheroidical form. A revolving fluid
will continue to change it's shape till it attains that in which
it's principles of contrary motion are balanced; for if you suppose them not balanced, it will change it's form. Now the same
balanced form is necessary for the preservation of a revolving
solid. The creator therefore of a revolving solid would make it
an oblate spheroid, that figure alone admitting a perfect equilibrium." Jefferson also concluded that the shape would "prevent a shifting of the axis of rotation."[8]

If God had created the earth, however, there was still a
major question: had he made it the way it is now, or in some
simpler pristine state? There was a host of unanswered questions about the present state of the earth's crust—formed into
mountains and valleys, folded, broken, with erosion of the
highlands and deposition in the lowlands. As Thomas Burnet
had observed after a visit to the European Alps a century before, "We must . . . be impartial where the Truth requires it,
and describe the Earth as it really is in its self . . . 'tis a broken
and confusd heap of bodies, plac'd in no order to one another;
nor with any correspondency or regularity of parts."[9]

A great deal came to depend on the answer, if one could
be found, to the long-standing question: how did fossils of
marine shelled animals come to be encased in rocks found
high up on mountains? The presence of those fossils suggested that change in the earth's surface had been massive.
In *Notes on the State of Virginia*, Jefferson listed the possibili-

ties: had the mountains been thrust up from some ancient seabed, had the seas receded, or even were these fossils really only artifacts created by some property of the rocks themselves? Jefferson calculated that there was not enough water in the atmosphere for Noah's Flood to have been universal or to cover high mountains and concluded ironically (and, strictly speaking, blasphemously), "There is a puzzle somewhere."

Jefferson had collected fossil shells himself in the Blue Ridge Mountains and was convinced that they were "real" shells. But after he had published *Notes* he became briefly intrigued by the idea, apparently fashionable in Paris during the time Jefferson was there and endorsed by none other than Voltaire, that fossils could, after all, grow in rocks like crystals. "It is now generally agreed that rock grows, and it seems that it grows in layers in every direction, as the branches of trees grow in all directions. Why seek further the solution of this phaenomenon? Every thing in nature decays. If it were not reproduced then by growth, there would be a chasm."[10]

The idea seems absurd to our modern minds, but it was perfectly familiar and accepted at the time. It offered an attractive way of "saving the phenomenon." It was given new legs by a Monsieur de la Sauvagiere living near Tours, who claimed actually to have observed fossils growing in rocks. When traveling in France in 1787, Jefferson followed up on the story, although the Marquis de Chastellux warned him that, while de la Sauvagiere was a man of truth and might be relied on for whatever facts he stated as of his own observation, "he was overcharged with imagination, which, in matters of opinion and theory, often led him beyond his facts."

Jefferson was tempted but not persuaded, musing:

If it be urged that this does not exclude the possibility of a like shell being produced by the passage of a fluid thro the pores of the circumjacent body, whether of earth, stone, or water; I answer that it is not within the usual oeconomy of nature to use two processes for one species of production. While I withold my assent however from this hypothesis, I must deny it to every other I have ever seen by which their authors pretend to account for the origin of shells in high places. Some of these are against the laws of nature and therefore impossible: and others are built on positions more difficult to assent to than that of de la Sauvagiere. They all suppose these shells to have covered submarine animals, and have then to answer the question How came they 15,000 feet above the level of the sea? and they answer it by demanding what cannot be conceded. One therefore who had rather have no opinion, than a false one, will suppose this question one of those beyond the investigation of human sagacity; or wait till further and fuller observations enable him to decide it.[11]

Jefferson was fascinated by petrified shells; he had seen what looked like assemblages of fossilized shells in the Blue Ridge near his home. He was even more fascinated by the remains of the American mastodon that in midcentury began to turn up; large numbers of teeth, tusks, and bones were found at Big Bone Lick (in present-day Kentucky) and in New York State. The New York specimens were exhibited by Jefferson's friend Charles Willson Peale at the American Museum, of which Jefferson was a director. When William Clark collected for him a large number of remains (of mastodons and other mammals) from Big Bone Lick, Jefferson kept one-third for himself, gave one-third to the American Philosophical Society, and sent one-third to the Royal Collections in Paris. (It was the French zoologist Cuvier who subsequently gave the beast the name "mastodon"; Jefferson always referred to it as a mammoth.) In 1799 Jefferson also described for the first time the "great-claw" (megalonyx), a giant fossil ground sloth. In every sense he can be thought of as a founder of American paleontology.[12]

But in his attitude toward all these fossils Jefferson showed an overwhelming conservatism, even reactionism, that trumped any science and that should have silenced his severest religious critics. In fact, he never actually referred to his mastodon and Megalonyx or great-claw "big bones" as "fossils." He correctly argued, against Buffon, that the mastodon and mammoth of northern climates were elephant species created for cold climates and different from the African and Asian elephants, but he never accepted the fact, or even the concept, of extinction. He argued fiercely again extinction in both *Notes on the State of Virginia* and in his 1799 paper on the Megalonyx.[13] In his view, all life was created by God and hence had been created perfect. Why, then, would God change his mind and bring in new sets of creations? In support of this, he relied also on travelers' (tall) tales from the American West to argue that these giant mammals were still alive, out there in the wastes toward the Great Stoney Mountains.[14]

Jefferson had summed up his philosophy in a few lines of his 1786 letter to Charles Thomson. "Theorists . . . must suppose a creator then; and that he possesed power and wisdom to a great degree. As he intended the earth for the habitation of animals and vegetables is it reasonable to suppose he made two jobs of his creation? That he first made a chaotic lump and set it into rotatory motion, and then waiting the millions of ages necessary to form itself, that when it had done this he stepped in a second time to create the animals and plants which were to inhabit it? As the hand of a creator is to be called in, it may as well be called in at one stage of the process as another. We may as well suppose he created the earth at once nearly in the state in which we see it, fit for the preservation of the beings he placed on it." The word *nearly* is important here and we must return to it in a moment.

Jefferson had written largely in response to Thomson's recommendation that he read a new work entitled *An Inquiry into the Original State and Formation of the Earth* by the self-taught British inventor, philosopher, and geologist John Whitehurst. Whitehurst had developed perhaps the last really innovative variant on the idea, popular in the eighteenth century (despite the fact the Bible records that mountains were present before Noah's Flood: Genesis 7:20), that the earth had originally been created smooth like an egg and that its current complex state was due to the Flood of Noah. Whitehurst's ingenious idea involved water leaking into the molten depths of the earth, where it was superheated to steam, causing vast explosions and throwing the whole crust into semi-chaos. Jefferson was unimpressed: "You ask me what I think of his book? I find in it many interesting facts brought together, and many ingenious commentaries on them. But there are great chasms in his facts, and consequently in his reasoning. These he fills up by suppositions which may be as reasonably denied as granted. A sceptical reader therefore, like myself, is left in the lurch." Thomson replied that he had not meant Jefferson to take Whitehurst too seriously, but "I think you will give him some credit for solving some of the objections started by the theorists against the universality of the deluge; and for accounting with a great deal of ingenuity for the present appearances and irregularities on the face of our globe." Indeed, Thomson suggested, "His eruption will tolerably well account for the oblique position of the strata of rocks which is observable in most parts of the world."

Thomson here brought up a confusing issue of American geology: "But what are we to think of their horizontal position in our Western country?"[15] Anyone exploring the Appalachian Mountains would have known the extent to which the strata there lay at all angles. The same explorer venturing westward into the land beyond the mountains also knew that the rocks out there were laid down in more or less perfect horizontal order, just as one would expect from a single divine Creation. Jefferson's reply (written almost a year after his earlier geological letter) showed that he had been thinking a lot about geology since writing *Notes*.

As he addressed directly the question of the origin and modification of mountains and geological strata, Jefferson revealed his knowledge of contemporary geological literature and practical geology and, yet again, his conservatism.

> With respect to the inclination of the strata of rocks, I had observed them between the Blue ridge and North Mountain in Virginia to be parallel with the pole of the earth. I observed the same thing in most instances in the Alps between Nice and Turin: but in returning along the precipices of the Appennines where they hang over the Mediterranean, their direction was totally different and various; and you mention that in our Western country they are horizontal. This variety proves they have not been formed by subsidence as some writers of theories of the earth have pretended, for then they should

always have been in circular strata, and concentric. [This is a reference to Nicolas Steno's 1669 *Dissertationis prodromus;* see below.] It proves too that they have not been formed by the rotation of the earth on it's axis, as might have been suspected had all these strata been parallel with that axis. [This is a reference to Buffon.] They may indeed have been thrown up by explosions, as Whitehurst supposes, or have been the effect of convulsions. But there can be no proof of the explosion, nor is it probable that convulsions have deformed every spot of the earth.[16]

In *Notes on the State of Virginia*, however, Jefferson had already shown his hand or at least had left the door open for a more modern theory of the earth. He loved to expound on the sublime landscapes of America (no doubt in response to those who extolled the superiority of the European Alps). In his chapter on mountains, he described one of his favorite places, the dramatic topography of the confluence of rivers at Harper's Ferry. The passage is worth quoting in full to show the passion that Jefferson felt about geology and landscapes. Jefferson noted that the Appalachian Mountains formed a series of southeast-northwest parallel ridges with great valleys between. These mountains were broken through at various places by rivers running east to the ocean.

> The passage of the Patowmac through the Blue ridge is perhaps one of the most stupendous scenes in nature. You stand on a very high point of land. On your right comes up the Shenandoah, having ranged along the foot of the mountain an hundred miles to seek a vent. On your left approaches the Patowmac, in quest of a passage also. In the moment of their junction they rush together against the mountain, rend it asunder, and pass off to the sea. The first glance of this scene hurries our senses into the opinion, that this earth has been

created in time, that the mountains were formed first, that the rivers began to flow afterwards, that in this place particularly they have been dammed up by the Blue ridge of mountains, and have formed an ocean which filled the whole valley; that continuing to rise they have at length broken over at this spot, and have torn the mountain down from its summit to its base. The piles of rock on each hand, but particularly on the Shenandoah, the evident marks of their disrupture and avulsion from their beds by the most powerful agents of nature, corroborate the impression. . . . This scene is worth a voyage across the Atlantic.[17]

There is one small phrase, buried in all this eloquence, that reveals the extent of Jefferson's dilemma over Creation and geology: "this earth has been created in time" (that is, over time). These published words were echoed in what he later wrote privately to Thomson about the world having been created "nearly as it is." Jefferson clearly allowed that Creation had not after all been a single event in which the world was formed just as it is now.

The matter sat largely unremarked until the election of 1800, which coincided with a broad Christian revival movement across the States. Then he was savagely attacked for being an "infidel" who denied the literal truth of the account of Creation in Genesis. It turned out that Jefferson had put himself on the wrong political side of a scientific argument. Clement Clarke Moore is known best to us today as the author of "A Visit from Saint Nicholas," or "'T'was the Night Before Christmas," a gentle children's story, a fantasy involving materially impossible characters and events. On the subject of Jefferson's geology, however, Moore (professor of Oriental and Greek literature at the General Theological Seminary in New York) was by turns sarcastic and vitriolic: "Whenever modern philosophers talk about mountains," Moore wittily observed, "something impious is likely to be at hand. . . . [Jefferson's geology] seems to posses every qualification which the heart of a modern philosopher could desire; it is bold, plausible, and contrary to Scripture." *Notes on the State of Virginia*, he said, tends "to the subversion of religion," and he asked "whether, from brilliancy of invention, acuteness of investigation, or cogency of argument, they [*Notes*] are entitled to the name of any other than modern French philosophy." Where Jefferson had been very careful to qualify his statements on the origins of mountains and to set them within the accepted biblical account of Creation, Moore read him as an out-and-out atheist. Moore said that Jefferson offered "a theory of the earth contrary to the scripture account of the creation . . . [that] denies the possibility of an universal deluge . . . [and] considers the Bible history no better than ordinary tradition." And behind this were "Voltaire and the French Encyclopedists, the imps who have inspired all the wickedness with which the world has of late years been infested."[18]

In the election of 1800, everything that Jefferson stood for—including science—rankled with the Federalists, perhaps not least because his philosophies seemed to favor two outcomes: the creation of an intellectual, philosophical elite and the concept of "progress" in American life and nationhood. Jefferson was ridiculed for his interest in mastodons, and *Notes on the State of Virginia* was characterized as nothing more than an exercise in measuring mice and rats and the genitalia of Indians.

It has to be said also that the Reverend Timothy Dwight, president of Yale College and one of the leaders of the religious revival, despised Voltaire and the French and, by association, Jefferson: "To what end shall we be connected with men, of who this is the character and conduct? . . . Is it, that we may see our wives and daughters the victims of legal prostitution; soberly dishonoured; speciously polluted; the outcasts of delicacy and virtue; and lothing of God and man?"[19] Electioneering was then no less brutal and unsophisticated than it is now.

Jefferson was in a perfect bind, having found his science and religion in conflict. He was not afraid of confronting authority. In his deism he denied the tenets of Christianity and had even gone so far as to produce his own version of the New Testament. But when a new geology tested his worldview, it created a step too far for him to take. The science was, as yet, not solid enough to throw away a lifetime's thought about religion. For all his devotion to science, he chose the religious answer over the scientific, but he never could quite close the door to the latter.

Modern commentators find Jefferson infuriatingly inconsistent. Joseph Ellis famously referred to him as a "sphinx."[20] Part of his mystery is due to his scientist's need for facts and truth. He held firmly to his truths, but those truths were sometimes in conflict. For example, with respect to slavery, he hated slavery but equally strongly believed that blacks could not become free citizens. He thought they might be expatriated to Africa or the West Indies, but then he also rejected the notion that European labor could replace them.

The historian Charles Miller, in his landmark *Jefferson and Nature*, even states that Jefferson had a "disrespect for geology." Miller decried Jefferson's "uninformed, impatient, and ultimately confusing discussion." He complained of the vacillation by which Jefferson sought to escape "the intellectual warfare by adopting, at least temporarily, the growth of fossil hypothesis."[21] I suggest, however, that Miller's interpretation is itself somewhat impatient. We cannot expect our eighteenth-century heroes (or indeed our villains) always to be so clear-headed that they had understood and correctly analyzed issues that took another two hundred plus years to resolve (assuming that we are right!).

In the case of geology, Jefferson saved the phenomenon of his religious views in the simplest way. He was baffled by geology and, seeing no way forward, decided to do nothing and go on with the rest of his life. But the intellectual world really was moving, so to speak, under Jefferson's feet, and in ways with which, one can only suspect, he must have sympathized. Meanwhile, he was, as we often are, damned if he did and damned if he didn't. His deism was not enough for the evangelical Christians, and he disappointed others by not taking his science far enough.

In the circumstances, what Jefferson did is what we all often do: he hung onto what he thought was right and waited things out. He didn't dash off in some silly direction (if we exclude the temporary aberration of the rock growth idea), nor did he lash out at one side or another, even if their ideas caused him pain.

As time went by, he adopted a somewhat rather peevish and even anti-intellectual approach to it all. He continued to argue that mineralogy and geology were central to a young person's education, but he rejected theory. In 1805, Jefferson wrote to the French immigrant anthropologist and geographer C. F. C. de Volney that he no longer "indulged . . . in geological inquiries." He had come to believe that "the skin-deep scratches which we can make or find on the surface of the earth, do not repay our time with as certain and useful deductions as our pursuits in some other branches."[22]

Jefferson made an even more negative statement about geological theory in 1826, in a letter advising Dr. John Emmett on a curriculum for the new University of Virginia: "The dreams about the modes of creation, inquiries whether our globe has been formed by the agency of fire or water, how many millions of years it has cost Vulcan or Neptune to produce what the fiat of the Creator would effect by a single act of will, is too idle to be worth a single hour of any man's life."[23] But while he, on the surface, rejected geology, his plan for the curriculum for his new University of Virginia included geology. Dismissive as he was of geological theorists, he kept up to date with what they were saying. In the end, one suspects, Jefferson would have preferred to keep his doubts about the geology of Creation private, but in writing *Notes* intellectual honesty required that he should describe what he saw and try to interpret its meaning, if ever so hesitantly, in public. There was, indeed, "a puzzle somewhere."

Perhaps the most telling lesson from Jefferson's dilemma over geology is how difficult it is for us to put ourselves into eighteenth-century shoes. What was obvious then now seems wrongheaded; what was baffling is now explained. And while, as in the following chapter, the tangled lines of inquiry and explanation seem eventually to align themselves, we can never be certain what different directions future knowledge, placed in a different context, might produce.

CHAPTER FOUR

Ancient of Days

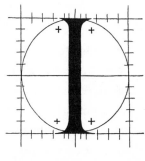N discussions about the relations between religion and science, we often forget that there was at least one important moment in history when the two came together. Ironically, it concerned no less an issue than Creation itself. By the time of Jefferson's death in 1826, geologists and philologists together had started completely to rewrite the early story of the earth for those who would read it. Whether Jefferson would have found their conclusions comforting or more confounding is hard to know.

These new generations of scholars challenged two basic notions. They said that the first verse of Genesis did not describe a single day but an unknowably long period during which the earth was created. The evidence for this was philological and depended on the use and meaning of the Hebrew words for "created" and "made," that is, out of nothing or out of something. (The first unequivocal reference to the earth having been created out of nothing seems to be in 2 Maccabees 7:29.) The second conclusion was simply commonsensical, if revolutionary: that the subsequent "days" of Creation were not actual days but figurative ones, each a long period. (And in doing so, among other things, these scholars gave new credence to St. Augustine's opinion in *The Literal Interpretation of Genesis* that the six days of Creation was not meant literally.) When Charles Darwin set off on HMS *Beagle* in 1831, Benjamin Silliman at Yale was teaching a new version of biblical geology, and (to the fury of critics) his ideas were published in one of the most popular geological textbooks of the time.

Theologians had argued among themselves for centuries about the wording and meaning of the first chapter of Genesis. In the latter half of the eighteenth century, just as people like Jefferson were trying to come to terms with geology and the Bible, both theologians and scientists attacked the problem. Among the elemental questions that theologians asked were: is Genesis a literal account of Creation? Is it (along with the rest of the Pentateuch) authentically the literary work of Moses himself? Was it dictated to Moses by God? Scientists asked of the earth itself: how was the earth formed initially? Had it been modified subsequently, and if so how and to what extent? What is the history of life on earth, especially in light of the fossil record? And both theologians and scientists, using their different methods and resources, tried to answer the perennial, deceptively simple question: How old is the earth?[1]

The history of these attempts by geologists and theologians (often that meant the same individuals) to extend and accommodate their understandings of what the discoveries of the Age of Reason meant for interpretation of the Book of Genesis cannot be told as a simple narrative. There was not one idea whose history can be traced, nor one set of influences. Rather, the story is a mosaic of events and people and ideas, which adds a further irony since the subjects under discussion were always, sooner or later, Moses and the "Mosaic" account of Creation in the first chapter of Genesis; of those, the following is a series of samples.

For centuries the twin propositions that the Pentateuch was literally a true history of the origin of the universe and of the inhabitants of earth and that the author of this account was none other than Moses himself were not worth the risk of challenging. Authority said it was so, and there was little advantage in disputing "authority" (in public, at least). Depending upon the age in which one lived, there were often quite a few reasons for not doing so.

From the third century onward, pious men like Origen had attempted to read the genealogies in the Pentateuch to discover two important dates: the day that the world was created and the day of the Flood of which Noah and his family were the sole human survivors. Their answers varied within a small range and by and large were not so very different from the dates famously promulgated by Bishop Ussher in 1650. The earth, according to the book of Genesis, was about six thousand years old, and the Flood happened around two thousand years after Creation. There were, of course, many who tried to improve on Ussher's dates; the world fondly remembers Sir John Lightfoot, vice-chancellor of Cambridge University, who decided that while Ussher gave the First Day as Sunday, October 23, 4004 BC, the exact moment had been at 9.00 a.m. (This embellishment is often attributed to Ussher.)

From at least the fifteenth century, however, actual observation of the earth's surface suggested that a Creation in six days was unlikely. The earth's crust was not simple and pristine. It was observably fractured, warped, and broken and, as Jefferson knew, to his distress, that bespoke change. If nothing else, it was obviously subject to change through erosion. In the late 1600s, both the English scientist Robert Hooke and the Danish anatomist (and later Catholic bishop) Nicolas Steno recognized that sedimentary rocks were formed by the erosion of uplands that were washed into the sea and became consolidated there into layered strata. So far, so good, but such layered strata were to be found in mountains, and that raised the question of how mountains came to be. Hooke suggested that they were raised up by confusions of the earth's surface like earthquakes and volcanoes, driven by the inner heat of the earth.

The discovery of fossils high up on the sides of mountains added to the question. As Jefferson had noted, it could only mean that either the mountains had been raised up from what had once been the sea floor or that the seas had dramatically receded. Perhaps an impossibly large flood had alerted the entire landscape. The only other possibility—that the fossils were only artifacts of the rocks themselves—was rendered moot when close study of fossil shells showed that they could be found in life positions and different life stages, just like living shells.

At the same time as Hooke was lecturing at the Royal Society in London on his ideas, Steno outlined a theory that involved the periodic collapse of an originally featureless earth into subterranean caverns, leaving high ground exposed in places as mountains and forming sedimentary basins in between. The high grounds were gradually eroded, building up sedimentary strata in those basins. There were three cycles of collapse and deposition. The last collapse accounted for the great Flood, in which "the waters of the deep opened."

Whatever the value of his theory of the earth, Steno made one magnificent contribution to the study of geology. He articulated three devastatingly simple principles: strata are always laid down horizontally; where layers of rock sit on top of each other, the higher ones are younger than the deeper ones and had been laid down independently of the lower ones; and single strata traced out laterally must all have been laid down at the same time.

Hooke's and Steno's cyclical theories did not even attempt to accommodate the biblical timeframe of a six-day Creation, and it is remarkable that Steno's was passed by the censors of the Inquisition just thirty-five years after Galileo, working under the same patron (Ferdinand II de Medici) had been convicted. Meanwhile, most seventeenth- and eighteenth-century theorists concentrated instead on finding ways of explaining any changes in the face of the earth in terms of some recent catastrophe like the Flood or, after Halley's discovery, an impact with a comet or comets that could, by a stretch of the imagination, be construed as being compatible with Genesis.

Scholars like Buffon followed a different track. Observations in mines showed that the earth was observably hotter inside than on the surface and therefore must be cooling. And this gave rise to theories that the earth had once been like the sun, a mass of superheated gas and molten rock. Perhaps it had once been a part of the sun, spun off by a collision with a comet. Also popular was the Nebular Hypothesis of Kant and Laplace, which posited that the solar system had begun as a hot cloud of gas and dust that collapsed under its own gravitational force and began spinning, gradually coalescing into a number of solid planets revolving around a central sun. Whatever the underlying process, cooling, mountain building, erosion, and sedimentation, faulting, and warping all required unimaginably long periods of time. However long that period had been, it must have been far longer than six days.

A crucial part of the mosaic of new discussions about Genesis came not from geology but from literary analysis. Jean Anstruc, physician in Montpelier and Paris and the author of the first authoritative work on venereal diseases, was a keen biblical scholar and linguist. In the previous century, Hobbes (*Leviathan*, 1651) and Spinoza (*Tractatus Theologico-Politicus*, 1670), among others, had newly questioned whether Moses had been the author of the Pentateuch. In 1753 Anstruc published anonymously a small book tackling the problem that had been there all along—the obvious fact that Genesis is not a unitary work. The Hebrew Bible and Septuagint were compilations from at least two separate documentary sources. There are, after all, two accounts of Creation and two accounts of the creation of the first woman. There are two names for God: Jahweh and Elohim. There are two versions of the Flood story. And by then everyone had recognized that

Deuteronomy, at least, had to be the work of some author other than Moses, if for no reason other than that it contains a description of his death. Ironically, Anstruc was entirely respectful of the idea that God had indeed directed Moses in assembling the Pentateuch out of ancient narratives. But the tiny little hole that he had created in the shield of orthodoxy soon became a gaping chasm.

Anstruc's literary analysis helped open the gates for a flood of literary and philological analyses of the biblical texts. In the early nineteenth century, a whole school of "rationalist" theologians and philosophers, including Wellhausen, Herder, Acosta, Strauss, Kurtz, Bunsen, and Feuerbach, largely concentrated in Germany, soon arose. The resulting "Documentary Hypothesis" eventually expanded to posit as many as four or five literary traditions, originally oral histories from different ages, for the first five books of the Bible, all written within the first millennium BC.

The principal theological issue raised by all this work was that anything tending to weaken the view that God, via Moses, was the author of the Pentateuch attacked its status as God's direct revelation. Many authors rationalized the situation (saved the phenomenon) by suggesting that what God had dictated to Moses had been deliberately framed in simple storylike forms, rather than as a scientific description, because men's minds were not yet ready for more sophisticated science. As we have become more sophisticated, God has allowed us gradually to refine the story of Creation. For example, the Reverend Baden Powell, professor of mathematics at Oxford, wrote in 1859, "They used simple language, as they adapted the common beliefs of the day,—Indeed, even had it been otherwise, no other language could have been intelligible to those they addressed."[2]

Rev. Powell had written his collection of essays, a review of the history of science and of interpretations of the biblical and geological record, over a number of years and published them just before Darwin launched his theory of natural selection. He saw a direct connection between advances in science and theology: "The actual *evidence*, and thence in some measure the interpretation, of theological truth will . . . take a different form at different periods, according to the existing character and state of advance of physical knowledge. . . . In a wider sense, physical philosophy, as cultivated in any particular age, will exercise an indirect and powerful influence over the general tone of thought and reasoning of that age, which will extend itself to other subjects not immediately physical. . . . In both ways the state of natural science will manifest effect directly and indirectly bearing on that of theology."[3]

Just as the German philosophers began to be read in Britain and America (Strauss and Feuerbach were translated into English by George Eliot in the 1840s), the ideas of Buffon and Georges Cuvier, his successor at the Jardin des Plantes in Paris, were beginning to gain traction. As early as 1777, Buffon had extended his theory of the earth along the lines that, after its molten origin, the history of the earth fell into six major epochs. In the first of these, "the earth and planets assumed their proper form." In the second epoch, the fluid earth cooled enough to consolidate. In the third epoch, the continents were covered with water and life began. The waters receded in the fourth epoch, giving dry land, and this was when volcanoes first erupted. In the fifth epoch, "the elephants, and other animals of the south, inhabited the northern regions. In the sixth epoch, the continents separated, and the seventh epoch was marked by the arrival of man and his influence on nature. Buffon had turned the six days of Creation into six long, perhaps unknowably long, ages.

Following Buffon, Cuvier, with his assistant Broigniart, began exploring the geological formations and fossils that were being turned up as excavations proceeded right in Paris. In essentially founding the science of stratigraphy based on Steno's principles, he discovered a whole succession of ancient worlds, layered one upon the other, and each populated by different patterns of (now fossilized) animals. But whereas Buffon's epochs merged more or less steadily into each other, Cuvier saw the world as having had a violent history, with periods of relative quiet interrupted by catastrophic episodes, local in extent, of floods, after which new species were formed. The last of the catastrophes was, of course, the episode of the Noachian Flood.

Perhaps Cuvier's greatest contribution to the debate was to confirm the order in which "Creation" appears in the fossil record. Those animals and plants that were (and still are) considered simpler and more lowly in the view of science were found earlier in the fossil record: fishes before mammals; humans appeared last of all. It was obvious to connect the dots—the geological sequence that Cuvier revealed matched the order of Creation in Genesis.

Meanwhile, the Scottish philosopher James Hutton had in 1785 produced his own theory of the earth, one that eventually came to stand as the foundation of modern geology. With Hooke, Hutton believed that the earth had a fiery, molten beginning. The forces elevating the land into hills and mountains were earthquakes and the "expansive powers" of the earth's inner heat. With Hooke, he succeeded brilliantly in showing that the earth had been formed in at least three cyclical phases of erosion and uplift.

One of Hutton's greatest insights was to insist that the processes that have changed the face of the earth in times past were simply the same that can be observed in action today, acting at the same rate and on the same scale (uniformitarianism). He had even thought that he might be able to find the age of the earth by measuring the rate of those processes. But he failed, famously concluding that he could not find evidence of the actual "beginning" and therefore he could demonstrate "no vestige of creation, no prospect of an end."[4] This didn't mean, however, that he thought there had not *been* a beginning.

A milestone in the process of bringing all these ideas together for a lay audience was raised, appropriately enough, by a minister who was also a mathematician. In 1804 the Reverend Thomas Chalmers (who would become a major force in the Free Church of Scotland) gave a lecture at Saint Andrews University, for which he was soon to be famous (and controversial, of course), in which he laid out the first version of what is known as the "Gap Theory." It offered a significant place where theology and science could be comfortable together. In a sense, science would rescue religion from the growing doubts about the age of the earth.[5]

Chalmers's solution to the problem of accommodating geological discoveries with Genesis was to reinterpret the wording of verses 1 to 3 of Genesis: "In the beginning God created the heaven and earth. And the earth was without form, and void; and darkness was upon the face of the deep. And the Spirit of God moved upon the face of the waters. And God said, 'let there be light': and there was light." This did not mean, Chalmers argued, that the universe, "without form, and void," was created on day one. Instead, an unknowable long period of Creation had passed before "the evening and the morning were the first day."

It was during this immensely long time that "the Spirit of God moved upon the face of the waters." During that first period, the "gap," the earth as we know it in all its geological complexity, developed by the sorts of processes Hooke, Steno, Buffon, Hutton, Cuvier, and many others had been proposing. It was a place where Creation might proceed, and only after all the groundwork had been done did God embark on his work of Creation on days two through six. And for Chalmers that meant literally six twenty-four-hour days.

To make this accommodation required a new view of an old problem. In Genesis, it was not until the third day (verses 9–12) the God created dry land and said, "Let the earth bring forth grass, the herb yielding the seed, and the fruit tree yielding fruit after his kind." If Chalmers was right, then dry land should have been created in "day" one. Even more difficult was that light was created at the beginning, but the sun, moon, and stars were not made until the fourth day. But where did the first light come from and how could grass and trees (created on the third day) have lived without the sun? The

solution was to posit that the earliest phase of Creation produced a general light that was obscured by deep clouds until
day four: an awkward compromise. Debate over the "light"
continued well into the second half of the nineteenth century.

In fact, the idea that an indefinite period had passed between the creation of matter and what was often referred to
as "the work of the six days" was an old one, dating back to
Justin Martyr, Gregory Nazianzen, Origen, and Augustine.
Nonetheless, Chalmers had opened a way for geology and
Genesis to be reconciled. All the time that was needed for
vast expanses of earth history was allowed by the vagueness of
the very first phase, when "In the beginning God created the
heaven and earth."[6]

Inevitably, however, if that infinitely long "beginning"
could be separated from what followed, the second obvious
question was: how long was a "day?" Was a day, as in "millen-
nial theory," for example, a thousand years, with the seventh
millennium being when Judgment would come? For geologists such as Jean-Andre Deluc in the late eighteenth century,
there was only one answer: everything discovered in the geological record could only have been produced by processes
acting over very long periods of time.

Benjamin Silliman at Yale took up the issue in terms of
the vastly increased knowledge of the 1830s, first in an appendix to the influential English textbook *Introduction to Geology* by Robert Bakewell, the first edition of which had appeared in England in 1813. Contributing an appendix to the
third (first American) edition of 1829 summarizing his Yale
lectures, Silliman was relatively circumspect: "The creation
of the planet was no doubt instantaneous, as regards to ma-

terials, but the arrangement, at least of the crust, was gradual. As a subject either of moral or physical contemplation, we can say nothing better, than that it was the pleasure of God that this world should be called in existence . . . but . . . that the arrangement . . . was to be progressive." And he showed that the progressive nature of the fossil record could be thought consistent with the progression of Creation in days two to six of Genesis, even to humans coming last.[7] "This is in strict analogy with the regular course of things in the physical, moral and intellectual world. Every thing, except God, has a beginning, and every thing else is progressive. It is of no importance to us, whether our home was in a course of preparation during days or ages, for the moral dispensation of God toward man could not begin until the creation of man. The gradual preparation of this planet for its ultimate destination presents therefore no anomaly, and need not excite our surprise."[8]

In 1833 (and then again in 1839 when his lectures were summarized as a separate book), Silliman revised and extended his views. In the final version, he extended his discussion from geology itself to an outright analysis of the theological consequences. As for "days," to the prior conclusion that "the arrangement, at least of the crust, was gradual," he added that critics think that the "period alluded to in the first verse of Genesis, 'in the beginning,' is not necessarily connected with the first day. It may therefore be regarded as standing by itself, and as such it is not limited, it admits of any extensions backward in time which the facts may require. . . . By asserting that there was a beginning, it is declared that the world is not eternal. . . . The world was, therefore, made in time by the omnipotent Creator." Then, finally, he admitted that "it does not appear to us necessary to limit the word day, in this account, to the period of twenty-four hours."[9] And thereafter he substituted the term "periods of time" for "days."

Silliman's revisionism did not immediately carry the day; he was fiercely opposed from the theological side by a former Yale colleague Moses Stuart (professor of sacred literature at Andover Seminary) and from science by the chemist Thomas Cooper (University of Pennsylvania and later South Carolina).[10] Stuart complained of his infidelity to the Bible while, oppositely, Cooper charged that Silliman's work represented an "unconditional surrender of his common sense to clerical orthodoxy."[11]

In his 1817 inaugural lecture at Oxford (as reader in geology) and his 1833 *Bridgewater Treatises* volume, the Reverend William Buckland also favored the Gap Theory. He drew his support for the concept from his colleague the Reverend Edward Bouverie Pusey, professor of Hebrew at Oxford, whom he quoted anonymously to the effect that correct translation of the Hebrew (*bara*—"create") allowed for the Gap Theory, although the theory was far from popular even with this clerical support.

Buckland's main geological interest was Pleistocene, and he found evidence for a massive flood in the excavation of valleys and deposits of vast beds of sand and gravel all over Europe. Buckland's counterpart at Cambridge, Adam Sedgwick, agreed, but there was already considerable opposition to the notion of a flood, at least one great enough to have covered all the world, including the Alps and the Andes. Such a worldwide flood would have to have been some six miles deep, and even allowing for the biblical "opening of the fountains of the deep," there was not enough water available. Then, in 1836, Sedgwick made a remarkable volte-face. He announced in his presidential address to the Geological Society of London that he saw no geological evidence for a universal flood. Soon thereafter, the young Swiss geologist Louis Agassiz demonstrated unequivocally that what looked like evidence for a flood was caused by something that acted more slowly and infinitely more powerfully—glaciers.

The "days" were lengthening just as, one by one, invocations of the Noachian Flood were losing their usefulness in reconciling geology and Creation. But once again a compromise was possible. In 1840, the English Congregationalist seminarian John Pye Smith wrote a work in which he, like many others, tried to "harmonize" the differences between what scientists and theologians thought about Creation: "I trust . . . that we [will] neither torture the Bible to make it speak the language of philosophy, nor suppress or mutilate the facts of nature in order to bring about an agreement with the Bible."[12]

Chalmers had become famous for having said, "The writings of Moses do not fix the Antiquity of the Globe." Pye Smith now concluded similarly that "it never entered into the purpose of Revelation to teach men geographical facts, or any other kind of physical knowledge."[13] And he added a new wrinkle to the discussion, arguing from the authority of the second account of Creation in Genesis (chapter 2), rather than the first. What was described in that second account evidently did not apply to the whole earth but only the Levant.

Indeed, the word *eretz*, translated as "earth," could not mean the whole globe but just *country* in the sense of a local region. While elsewhere there was light, over the Middle East there were clouds and darkness until later. What Genesis described as Eden was not the whole world, and Creation happened there on a different timetable from the rest of the world. The Flood was then not a worldwide event but local to the Levant, and thus it was possible for Noah's sons to find wives and fulfill God's commandment to "be fruitful, and increase, and fill the earth" when all other humans had, in theory, been killed.

> I must profess then my conviction that we are not obliged by the terms made use of, to extend the narrative of the six days to a wider application than this; a description, in expressions adapted to the ideas and capacities of mankind in the earliest ages, of a series of operations by which the Being of omniscient wisdom and goodness adjusted and furnished the earth generally. But as the particular subject under consideration here, a PORTION of its surface, for most glorious purposes I conceive to have been a part of Asia, lying between the Caucasian Ridge, the Caspian Sea, and Tartary, on the north, the Persian and Indian Seas on the south, and the high mountain ridges which run at considerable distances, on the eastern and the western flank. . . . This state was produced by subsidence of the region . . . overflowed with water, and its atmosphere so turbid that extreme gloominess prevailed and "darkness was upon the face of the deep." The atmosphere then cleared, elevations of the land took place and consequently the waters flowed into the lower parts, producing lakes, and probably the Caspian Sea. . . . The elevated land was now clothed with vegetation instantly created.[14]

For many, the work of further harmonizing geology and theology meant finding some kind of direct correlation between the events described in Genesis with the fossil record, and that meant accepting both the Gap Theory and the "long" interpretation of a "day." Three writers in particular are worth singling out as the most widely read and influential writers on geology on either side of the Atlantic, brilliantly riding the popular interest in the subject of the 1850s and 1860s. In Scotland, there was the self-taught polymath Hugh Miller, a man who started out as a stonemason and became a paleontologist, journalist, and author. In America, the Reverend Edward Hitchcock, a student of Silliman who had been a chemist, geologist, and third president of Amherst College was joined by his son Charles Hitchcock, professor of geology at Dartmouth College. These three could scarcely have been more different, the professors and the journalist, but all had a firm grasp of the facts of paleontology and geology and a drive to make them conformable with the Bible.

This was a new golden age of geology. A crucial step had been made in the 1790s when the English canal surveyor William Smith, following the principle that Steno had articulated so long before, had the insight that each geological stratum had its own individual signature of lithology and fossils that made it possible to follow the formation to be traced over many miles, even if the surface exposures were discontinuous. This allowed the drawing of the first extensive geological maps. Then, between 1800 and the 1830s, knowledge of the fossil record began the exponential increase we have seen ever since. How exciting it must have been in the nineteenth century to see the weird and wonderful mammals from the Paris Basin, and the first dinosaurs, ichthyosaurs, pterosaurs, plesiosaurs, and Triassic mammals from prosaic English quarries and cliff sides!

Charles Hitchcock continued to promote the idea that Moses (God) had used the words "days" because men were not ready to understand more.[15] Hugh Miller disagreed with Pye Smith about a "local" rather than universal Flood but, having first held to the interpretation of "day" as a literal twenty-four-hour period, he reluctantly admitted that the geological record showed incontrovertibly that six days of Creation must have been long ages. In a series of extraordinarily popular books with evocative titles like *Testimony of the Rocks* and *Footprints of the Creator*, Miller used his vast knowledge of the fossils that were beginning to be found in such large numbers, and his gift of writing, to argue passionately that the geological record was still compatible with scripture. For both the Hitchcocks and Miller, the key issue was what Cuvier had outlined: the progression of life seen in the fossil record roughly matches the order of Creation in Genesis. By this time, the basic structure of the geological column had been worked out. Soon a minor cottage industry developed in trying to match up the six days of Creation (whether real days or long periods) with the elements of the geological column. For most, the task of the geologist was now principally to explain the meaning of just three days of organic Creation. And by taking a certain literary license with Genesis, this was possible. In his 1867 book, for example, Charles Hitchcock produced a detailed chart matching the days to geology and fossils. Four main geological eras had been identified: Azoic (days one to three—here Hitchcock included the first plants and protozoans); Paleozoic (day four: plants, fishes, amphibians); Mesozoic (day five: birds, reptiles, small mammals); and Cenozoic (day six: modern life, large mammals, and man). And gratifyingly, it remained true, for the moment, that no fossil record of humans existed.

Serious difficulties remained if one tried to take things too literally, for example if the sun, moon, and stars really were created on day four. But Hitchcock and Miller had produced a very good compromise—for the time being.

One of Miller's additional contributions was to promote a way in which, in the enlarged scale of days, the Sabbath day when God rested could be interpreted. God's last act of Creation produced humans, "and with the human, heavenly aspiring dynasty, the moral government of God, in its connection with at least the world which we inhabit, 'took beginning.' And then creation ceased. Why? Simply because God's moral government had begun . . . mere *acts of creation* could no longer carry on the elevatory process. The work analogous in its end and object to those *acts of creation*, which gave our planet its successive dynasties of higher and yet higher existences, is the work of REDEMPTION. It is the elevatory process of the present time, the only possible provision for that final act of *re*-creation 'to everlasting life,' which shall usher in the terminal dynasty."[16]

When Darwinism arrived, Miller was one of the most influential of those who argued that, even in its expanded, multimillion-year form, the geological record was a sure sign of the hand of the Creator. He made geology a key to an antievolution argument. Just as Cuvier had argued against a Lamarckian development theory, Miller argued passionately that there had been no gradual evolution of the sort that

Darwin was currently proposing. The apparent "progress" over time among all species was actually a parade of discrete divine creations, not a continuous material process of change. The only "change" had been in that with which God chose sequentially to populate the world. The essential nature of the fossil record was its discrete nature; there were no transitions between strata. Some of this was due to a circularity of argument produced by Smith's stratigraphy, which depended on defining strata on the basis of discrete internal similarities in their fossils. But Darwin himself had admitted the problem of gaps in the fossil record in his remarkable chapter "Difficulties with the Theory" in *On the Origin of Species*.

Nonetheless, a wide door had been opened; theories of change like Darwin's evolution by natural selection could not "work" at all unless the world were extremely old, and gradual progressive evolution was more compatible with modern geological uniformitarianism than religious catastrophism. By the time Darwin returned from the voyage of HMS *Beagle*, in scientific terms most of the old certainties about Genesis and Creation had been overturned. From the labors of scientists, philologists, and theologians, a new synthesis of ideas and facts, still uneasy and controversial, to be sure, had provided a foundation for a way forward. And they had created a new basis for a confrontation between private doubt and authority.

Parenthetically, it would be wonderful to know how much of the philological and theological debate over Genesis Jefferson had followed. The odd thing is that this sort of textual analysis was the kind of work that would normally have excited Jefferson's attention. Words, language, rhetoric, argument were his bread and butter all his life. And he had, after all, not shied away from editing the four Gospels. The only clue we have is that in correspondence with John Adams about Goethe's thesis that the Ten Commandments in Exodus were not the originals that God wrote on tablets of stone, he wrote, "But the whole history of these books is so defective and doubtful that it seems vain to attempt minute enquiry into it; and such tricks have been plaid with their text, and with the texts of other books relating to them, that we have a right, from that cause, to entertain much doubt what parts of them are genuine."[17]

Just as Jefferson could not fully accept the consequence of an earth "created in time," the difference between the scientific view of geological origins and the biblical one has today never been accepted by everyone. Those whose faith made them adhere to the concept of the literal truth of Genesis were, and are, unpersuaded. Indeed, just when at least a working majority of late nineteenth-century scientists had come to accept an essentially modern view of the fossil record and the great age of the earth, the whole subject of the three literal days of Creation was brought up again by no less a person than the British prime minister William Ewert Gladstone. A formidable classical scholar, Gladstone followed Hitchcock and Miller in trying to match the order of origins in Genesis with the fossil record. He simplified greatly on the basis of the Septuagint, so that on the fifth day the "water-populations" and "air-populations" were created, followed by "the land-population" and man. And he emphasized a fact that had been there all along: if derived from a folk tradition, the order of Creation in Genesis is remarkably perceptive in that *gener-*

ally it puts the primitive before the advanced—for example, the fishes before the land animals. As Gladstone asked: "How came the author of the first chapter of Genesis, to know that order, to possess knowledge which natural science has only within the present century for the first time dug out of the bowels of the earth? It is surely impossible to avoid the conclusion that . . . his knowledge was divine."[18]

He framed all this as a public attack on Darwin's close friend Thomas Henry Huxley, who politely pointed out the impossibility of the assumptions in Gladstone's analysis. Their debate could easily have been conducted fifty years earlier or today.

CHAPTER FIVE

Mr. Darwin's Religion

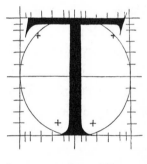

HOMAS Jefferson, when confronted with inconclusive contemporary theories of geology, essentially ducked the issue because he could not reconcile them with his very personal religious view of the Creation. In Charles Darwin's equally problematic case, the science was of his own making. Thomas Jefferson was quintessentially a man of, and for, the Age of Reason. He needed to be right, and he often was. But solid truths on which to base a worldview were sometimes hard to find, and science did not always provide the answers. He could be depressed and anxious—perhaps more than most of us, but his mental turmoil was nothing like that of Charles Darwin. Darwin suffered from chronic anxiety all his life. His famous illnesses (traced every year or so to another new and ever-more fashionable cause, such as mitochondrial disease) were based in that awful anxiety. Working on his theory of evolution by natural selection almost killed him, or at least he thought it would. A large part of the cause of this was the conflict between religion and evolution.[1]

A familiar theme of these essays is that important ideas, even when presented as new, tend to have long histories. Certainly evolution—usually first known as *transmutation theory* or *development theory*—was in the air long before Charles Darwin was even born. It was made familiar in philosophical circles at the turn of the century by Erasmus Darwin (Charles Darwin's grandfather) and the French zoologist Jean Baptiste de Lamarck. It became quite popular after 1844 due to the work of Robert Chambers, whose *Vestiges of the Natural History of Creation*, with its message of change, became popular among social reformers. When Tennyson wrote in *In Memoriam* (1849) of "nature red in tooth and claw" and, even more interestingly, "Are God and Nature then at strife, / That Nature lends such evil dreams? / So careful of the type she seems, / So careless of the single life," which sounds so Darwinian, he was repeating a common idea, captured by Thomas Malthus and, for instance, the French botanist de Candolle, that "all nature is at war with itself."

One of my favorite pre-Darwinian references comes, of all people, from Benjamin Disraeli: "You know, all is development. The principle is perpetually going on. First, there was nothing, then there was something; then—I forget the next—I think there were shells, then fishes; then we came—let me see—did we come next? Never mind that; we came at last. And at the next change there will be something very superior to us—something with wings. Ah! That's it: we were fishes, and I believe we shall be crows."[2]

Both Jefferson and Darwin could be remarkable writers. Unlike so many of their age, they wrote simply and with great clarity. For a Jeffersonian example we need look no further than the Declaration of Independence of 1776, although any of his writing already quoted shows the same thing. Darwin's *On the Origin of Species* can easily be given to writing composition classes for close study, 150 years after its publication, as an example of simple, elegant scientific prose. Most of what Darwin wrote was matter-of-fact expository prose, but a few passages are quite lyrical. Probably everyone's first choice for his best writing is the last paragraph of the first edition (1859) of *On the Origin of Species*, and particularly its last sentence.

It is interesting to contemplate an entangled bank, clothed with many plants of many kinds, with birds singing on the bushes, with various insects flitting about, and with worms crawling through the damp earth, and to reflect that these elaborately constructed forms, so different from each other, and dependent on each other in so complex a manner, have all been produced by laws acting around us. These laws, taken in the largest sense, being Growth with Reproduction; Inheritance which is almost implied by reproduction; Variability from the indirect and direct action of the external conditions of life, and from use and disuse; a Ratio of Increase so high as to lead to a Struggle for Life, and as a consequence to Natural Selection, entailing Divergence of Character and the Extinction of less-improved forms. Thus, from the war of nature, from famine and death, the most exalted object which we are capable of conceiving, namely, the production of the higher animals, directly follows. *There is grandeur in this view of life, with its several powers, having been originally breathed into a few forms or into one; and that, whilst this planet has gone cycling on according to the fixed law of gravity, from so simple a beginning endless forms most beautiful and most wonderful have been, and are being, evolved* [emphasis added].[3]

One obvious rhetorical purpose of this concluding paragraph was to end the book on a lyrical and supremely upbeat note because the theory of natural selection depends basically on the grim statistics and relentless arithmetic of the Malthusian world. Survival of favored races in the struggle for life produces change only if the unfavored races die. Direct and indirect struggle, competition for resources, disease, and want all drive natural selection. Nature is full of ugliness as well as beauty. What Darwin wrote is a direct refutation of the rosy teleology of Paley's *Natural Theology* and the utilitarianism of John Ray's earlier *The Wisdom of God Manifest in the Works of His Creation* (*1691*).

In the early outline of his theory that is usually called the *Essay* of 1844, the language Darwin used for this ending, while obviously similar, was even stronger.

It accords with what we know of the laws impressed by the Creator on matter that the production and extinction of forms should, like the birth and death of individuals, be the result of secondary means. It is derogatory that the Creator of countless Universes should have made by individual acts of His will the myriads of creeping parasites and worms, which since the earliest dawn of life have swarmed over the land and in the depths of the ocean. We cease to be astonished that a group of animals should have been formed to lay their eggs in the bowels and flesh of other sensitive beings; that some animals should live by and even delight in cruelty; that animals should be led away by false instincts; that annually there should be an incalculable waste of the pollen, eggs and immature beings; for we see in all this the inevitable consequences of one great law, of the multiplication of organic beings not created immutable. From death, famine, and the struggle for existence, we see that the most exalted end which we are capable of conceiving, namely, the creation of the higher animals, has directly proceeded. Doubtless, our first impression is to disbelieve that any secondary law could produce infinitely numerous organic beings, each characterized by the most exquisite workmanship and widely extended adaptations: it at first accords better with our faculties to suppose that each required the fiat of a Creator. There is a [simple] grandeur in this view of life with its several powers of growth, reproduction and of sensation, having been originally breathed into matter under a few forms, perhaps into only one and that whilst this planet has gone cycling onwards according to the fixed laws of gravity and whilst land and water have gone on replacing each other—that from so simple an origin, through the selection of infinitesimal varieties, endless forms most beautiful and most wonderful have been evolved.[4]

It is not always appreciated that *On the Origin of Species* was more than an exposition of Darwin's new theory. He had to rebut the conventional views of nature. And it was not just natural theology that had to be swept out of the way. Darwin also had to make a frontal attack on the prevailing Creationist view that each species was created independently by God through a divine spontaneous generation—that the entirety of all living and (presumably) fossil species had been created ex nihilo. As Darwin said in his concluding passages, "Several eminent naturalists . . . seem no more startled at a miraculous act of creation than at an ordinary birth. But do they really believe that at innumerable periods in the earth's history certain elemental atoms have been commanded suddenly to flash into living tissues?"[5]

And a few pages later he added, "Authors of the highest eminence seem to be fully satisfied with the view that each species has been independently created. To my mind it accords better with what we know of the laws impressed on matter by the Creator, that the production and extinction of the past and present inhabitants of the world should have been due to secondary causes, like those determining the birth and death of the individual. When I view all beings not as special creations, but as the lineal descendants of some few beings which lived long before the first bed of the Silurian system was deposited, they seem to me to become ennobled."[6]

Thus the "grandeur" of Darwin's "view of life" not only added a simple, elegant logic that preserves and even reinforces our sense of wonder at it all; he was proposing a new kind of grace—secular and ennobling.

In none of these passages did Darwin deny the existence of a Creator, but he was quite clear that the only role of the Creator had been to set in place *the laws of matter.* This was essentially a deist position in which the Creator did not guide the course of evolution, and he left it to the reader to imagine whether God was involved in the actual creation of the first life or whether that simply happened through the operation of material laws in some ancient molecules. The words "originally breathed," then, could be read figuratively or literally. These intriguing words then raise the question: what was Darwin's religion?

Darwin and religion had a constantly changing—an evolving—history. A very insular, brooding child, particularly after the death of his mother, he loved reading and lonely thoughtful walks. Brought up by his older sisters, he was devoted to his brother Erasmus but probably saw little of his father. He disliked being told what to do or think. As a sixteen-year-old student of medicine at Edinburgh, Darwin was the second-most prolific borrower of books from the university library. (Erasmus, with whom he roomed for the first year, borrowed the most.)

Darwin was brought up in a curious but at that time not uncommon combination of Unitarianism and the Church of England. Unitarianism, like the Baptist and Congregationalist churches, was then growing in popularity, especially outside of London in the industrial heartland where men tended to be "self-made" and both men and women thought for themselves. It was the religion of his Wedgwood kin, including his mother, and also of his deist grandfather Erasmus Darwin. Robert Darwin, Charles Darwin's father, joined the Church of England, as was fashionable for the upwardly mobile. Darwin was christened at the parish church in Shrewsbury but never confirmed. His wife-to-be, Emma Wedgwood, also an ardent Unitarian, was confirmed in the Church of England at age sixteen.[7]

Within Unitarianism a broad range of views was tolerated, centered on denial of the Trinity but allowing belief or disbelief with respect to the immortality of the soul and the divinity of Jesus. Miracles were not part of Unitarians' beliefs. One dominant feature of the Unitarians was (and still is) their insistence that discovery of the truth about God and

Jesus must come through personal study and reflection rather than dogma. This was the approach that Jefferson also always followed and, as for Jefferson, intelligent reflection was something at which Charles Darwin and Emma Wedgwood excelled. Both seriously revised their religious views as the years went by.

How much of the Unitarian's religious belief Darwin's mother had inculcated in him is something to be questioned and refined. And how much he, after her death and as a teenager, accepted, tolerated, or rejected the formal dogma of the Church of England, we will never know. It is possible even that his mother died too early (he was eight) to have any lasting effect on his religious development. His older sisters, however, were always quite religious and constantly tried to "improve" their younger brother in this regard.

As a young man, really still a boy, entering medical school at Edinburgh (1825), Darwin seems to have had very little idea of the formalities of religion. He and his brother Erasmus together visited a number of Edinburgh churches hoping to be entertained by some fire-and-brimstone preaching, but he had to write to his sister a sad little question, "What part of the Bible do you like best? I like the Gospels. Do you know which of them is generally reckoned to be the best?"[8]

As I have described elsewhere, Darwin's decision to leave Edinburgh and give up medicine was a complex one involving what he felt was betrayal by his teacher Robert Grant.[9] But it was the right thing to do. His next move, allowing his father (he had not much choice in the matter) to send him to Cambridge in 1828 to study for the church—seems almost cynical, especially as father and son had so little in the way of religious inclination, let alone any sense of a religious vocation. But the young man had to have a career of some sort, so Darwin crammed his neglected Greek in order to matriculate. He also had to convince himself that he could swear to the Thirty-nine Articles of the Church of England, another requirement for matriculation at Cambridge, where he entered Christ's College in January 1829 to start on the path toward becoming a country parson.

Darwin's problem, if we can call it that, was that he had not yet discovered something that engaged him fully as an intellectual. But it was an intellectual that he was striving to be. Darwin was always a quick study and a master of details. By the time he finished Cambridge he had at last became intimately familiar with the Bible. "Whilst on board the *Beagle* I was quite orthodox, and I remember being heartily laughed at by several of the officers (though themselves orthodox) for quoting the Bible as an unanswerable authority on some point of morality. I suppose it was the novelty of the argument that amused them." This was written in the 1879 *Autobiography;* the passage continues, "But I had gradually come, by this time, to see that the Old Testament from its manifestly false history of the world, with the Tower of Babel, the rainbow as a sign, etc., etc., and from its attributing to God the feelings of a revengeful tyrant, was no more to be trusted than the sacred books of the Hindoos, or the beliefs of any barbarian."[10]

Nothing in Darwin's letters and diaries from HMS *Beagle* suggests that he was looking forward with enthusiasm or pleasure to a career in the church; it would be nothing more than a social convenience of a kind only too familiar in England of the 1830s. At worst, he could become a gentleman naturalist in a comfortable "living" with a curate to do the daily work. For example, his cousin William Darwin Fox became rector of a church in Cheshire, and the prominent zoologist Leonard Jenyns, who was the brother-in-law of his Cambridge mentor John Stevens Henslow, was vicar of nearby Swaffham Fulbeck, Cambridgeshire. At best, with luck, ambition, and the right patronage, he might became a don at Cambridge like Henslow.[11]

The details of Darwin's transformation from naturalist to natural philosopher are well known. Whatever expectations he might ever have had concerning a church career began to wane even before the *Beagle* voyage began because he knew that he was going to be a wealthy man from his mother's Wedgwood legacy (received when he reached twenty-five) and his father's not inconsiderable fortune (several million pounds in modern terms). Expectations were totally lost by the time the ship returned to England.

As an undergraduate at Cambridge Darwin had to study William Paley's *A View of the Evidences of Christianity*; it was one of the set books he was examined on. *Evidences* did not expound abstract theological principles; rather, it examined the lives of the early Christian martyrs and concluded that a faith as strong as theirs must be authentic. At the time Darwin accepted Paley's argument completely. After his studies were complete he also read Paley's *Natural Theology* over and over and again, finding the argument compelling.

At Edinburgh Darwin had been taught a rather conservative kind of geology by Robert Jameson, a devotee of Cuvier. Cuvier, following the pathbreaking mineralogist Abraham Gottlob Werner at Freiberg, argued that the earth had been formed first in water (the Neptunist approach) and had then been shaped through a number of major catastrophes. At Cambridge Darwin spent less time on geology (or study in general) and instead dived into the contemporary fashion for insect collecting with an energy and skill that gained the admiration of the country's two greatest entomologists—the Reverend Frederick William Hope and James Francis Stephens. He exchanged specimens with both and while still an undergraduate he had no fewer than eleven entries in Stephens's encyclopedic *Illustrations of British Entomology* (1827–46). But two things were missing from his life (three, if we include the fairer sex). He had no ambition or prospects when it came to a career, and his questioning mind was not fully satisfied with simply collecting and identifying insects, even when he found new records for the British Isles. He may, however, have been somewhat seduced by seeing his name in print in Stephens's volumes. Then came the *Beagle* voyage.

During the *Beagle* voyage, it was geology that first led Darwin to question what he had previously learned from his clerical teachers at Cambridge and his family. His voyage around the world, with Charles Lyell's new and revolutionary *Principles of Geology* in his hand, settled matters. Once he had delved deeply into geology, any belief in the literal truth of the Pentateuch was impossible.

Very early in the voyage, the first landfall of the ship gave him the opportunity to test Lyell's theories against some real rocks. From that point he was fully engaged intellectually. Geology satisfied several drives—it involved hand specimens to be collected, identified, and classified. It involved theory of a broad and sweeping kind. And, finally, since Lyell's geology was a system of the whole earth, it set him on a path of investigating not just isolated phenomena—here and there, wherever an interesting situation turned up—but one grand system. That system was the long-term evolution of the earth's surface according to observable processes.

It was one thing to have read in books that the earth was variously broken or folded, the land raised or submerged, with landscapes slowly eroded and new sediments steadily built up, volcanoes and earthquakes altering the earth in a matter of minutes and hours—it was quite different to see it firsthand. He saw it all for himself and explored the earth in detail, from the coastal plains of South America to the high Andes and the coral reefs of the Indian Ocean. Everyone had read about the great Lisbon earthquake of 1755. At Concepción, Chile, in 1835 he experienced a massive event for himself. His officer friends on HMS *Beagle* were all surveyors; they measured, on the spot, that the land had been raised some three meters.

On the Galapagos Islands he saw where new volcanic land had been thrust up from the sea, and fields of lava lay about, seemingly newly congealed. For Darwin, the changing earth was not some abstract thing reported from classical authors; it was real. And it was perfectly clear to him that the earth must be much older than biblical arithmetic predicted, and that the vast changes he observed had not been caused by Noah's Flood. And soon his study of biological diversity led him in the same direction. In January 1838 he wrote in his notebook that the diversity of species on earth "leads you to believe the world older than *geologists* think."[12]

It was as a geologist, not primarily a student of animals

and plants, that he came back from the voyage. Within a year he was made secretary of the Geological Society of London. His first published scientific papers were geological in nature. Most important, one was also theoretical. That work concerned the origin of coral islands—one of the classic conundrums in geology—to which Darwin eventually devoted a whole book in 1844. This paper helped confirm Darwin's position as a geologist and intellectual. In fact, he had worked out his theory three years before the *Beagle* found its way to the reefs and atolls of the Indian Ocean where he put his ideas into practice. He read Lyell's ideas on the subject, saw that they must in principle be wrong, and devised the new theory entirely by himself. Any thought that careless commentators might have to the effect that Darwin was just an amiable but aimless young man before the *Beagle* voyage —in that case why was he taken along?—vanishes before this single contribution (which Lyell immediately accepted as superior to his own).

In the two years after his return from the voyage, Darwin's life changed irrevocably; he started to think in secret about the real possibility of transmutation of species, and by the end of 1838 he had set in place the last logical element of a theory of natural selection. And it was just then that he married his cousin Emma Wedgwood. Emma Wedgwood was a devout Christian, and one can only guess at the conflicts that arose between the direction his theory was taking him and the fundamentally (so to speak) opposite convictions of his wife. They continued to attend church together, and on the surface, for the moment, all was calm. Under the surface, however, as Darwin later admitted, he felt as if he were "confessing to a murder." If ever there was a case that demonstrated how difficult it can be to accept new facts, concepts, and theories and place them intellectually and personally amid the old, and how intensely it begins at the personal level, and if ever there was, in J. A. Thomson's terms, "a tax on new knowledge," this was it. It nearly killed him, or at least in his hypochondriacal state of mind he often thought it would. During his life he suffered not merely from symptoms like nausea but hysterical crying, feeling a sensation of walking on air, ringing of the ears, exhaustion, self-loathing, and "dying sensations."[13]

During these two years Darwin's doubts about the literal truth of the Bible and the basic tenets of Christianity surged. By the time of his engagement to Emma (November 11, 1838) and their marriage (January 29, 1839), not only had Darwin rejected the account of Creation in Genesis, his misgivings (to say the least) about conventional Christianity were also well set. Against the advice of his father, he opened his heart to his future wife about these matters, of which, he knew, she cared so very deeply. "Before I was engaged to be married, my father advised me to conceal carefully my doubts, for he said that he had known extreme misery thus caused with married persons. Things went on pretty well until the wife or husband became out of health, and then some women suffered miserably by doubting about the salvation of their husbands, thus making them likewise to suffer. My father added that he had known during his whole long life only three women who were skeptics."[14]

Darwin, always in a state of high anxiety, with too many secrets to keep, could not envisage a less than transparent beginning to the marriage, however. His letters to Emma are not known to be extant, but two of hers to him survive and they make clear that she respected his doubts and their basis in careful thought and analysis—but that they pained her greatly. Ten days after their engagement, Emma wrote to Darwin:

> When I am with you I think all melancholy thoughts keep out of my head but since you are gone some sad ones have forced themselves in, of fear that our opinions on the most important subject should differ widely. My reason tells me that honest & conscientious doubts cannot be a sin, but I feel it would be a painful void between us. I thank you from my heart for your openness with me & I should dread the feeling that you were concealing your opinions from the fear of giving me pain. It is perhaps foolish of me to say this much but my own dear Charley we now do belong to each other & I cannot help being open with you. Will you do me a favour? yes I am sure you will, it is to read our Saviours farewell discourse to his disciples which begins at the end of the 13th Chap of John. It is so full of love to them & devotion & every beautiful feeling. It is the part of the New Testament I love best. This is a whim of mine it would give me great pleasure, though I can hardly tell why I don't wish you to give me your opinion about it.[15]

Darwin's loss of faith is famous and has been endlessly argued over, although the very word *loss* is an awkward one as it begs the question of what Darwin's faith had ever been, and the extent to which he lost it. We have only tantalizing glimpses of the evolving course of Darwin's religious feelings. His 1879 *Autobiography* was written primarily for a family audience, and as he was often sure he was dying, he must have had a strong expectation that Emma would be around to read it after he died. So it was carefully worded. Evidently, however, during the decade between his marriage to Emma in 1839 and the death of his father (1848), Darwin's doubt and misgiving evolved into frank disbelief of conventional Christianity, and he states that he finally rejected all conventional Christianity at the time of his father's death. He had long been disinclined to believe threats of fire and brimstone and promises of heaven. In perhaps an unconscious play on words (Darwin was not much of a humorist, especially on a serious subject like this), he wrote that "I can indeed hardly see how anyone ought to wish Christianity to be true; for if so the plain language of the text seems to show that the men who do not believe, and this would include my Father, Brother and almost all my best friends, will be everlastingly punished. And this is a *damnable* doctrine" (emphasis added).[16] This was one of the many passages that Emma edited out of the first published version of the *Autobiography*. "I should dislike the passage in brackets to be published. It seems to me raw. Nothing can be said too severe upon the doctrine of everlasting punishment for disbelief—but very few now wd. call that 'Christianity,' (tho' the words are there.)"[17]

The final blow for Darwin came in 1851 with the death of his beloved daughter Annie at the age of ten. Darwin had long been examining over and over the tenets of natural theology and its failure to explain evil, misery, disease, and death as God's purpose. Just as the Concepción earthquake had literally shaken the ground beneath his feet, the death of Annie knocked away the last vestiges of Christianity for Darwin. "Thus disbelief crept over me at a very slow rate, but was at last complete. The rate was so slow that I felt no distress, and have never since doubted even for a single second that my conclusion was correct."[18]

There was, however, plenty of room for honest personal

doubt and, although he rejected all the trappings of Christianity, Darwin still wanted to believe in the existence of God. One good argument, he thought,

> connected with the reason and not with the feelings, impresses me as having much more weight. This follows from the extreme difficulty or rather impossibility of conceiving this immense and wonderful universe, including man with his capacity of looking far backwards and far into futurity, as the result of blind chance or necessity. When thus reflecting I feel compelled to look to a First Cause having an intelligent mind in some degree analogous to that of man; and I deserve to be called a Theist. This conclusion was strong in my mind about the time, as far as I can remember, when I wrote the *Origin of Species;* and it is since that time that it has very gradually with many fluctuations become weaker. But then arises the doubt— can the mind of man, which has, as I fully believe, been developed from a mind as low as that possessed by the lowest animal, be trusted when it draws such grand conclusions? May not these be the result of the connection between cause and effect which strikes us as a necessary one, but probably depends merely on inherited experience? Nor must we overlook the probability of the constant inculcation in a belief in God on the minds of children producing so strong and perhaps an inherited effect on their brains not yet fully developed, that it would be as difficult for them to throw off their belief in God, as for a monkey to throw off its instinctive fear and hatred of a snake.[19]

When he set off on the *Beagle* voyage at the age of twenty-two, these doubts had not occurred to him. Nor did he question the orthodox view of Creation that the observable patterns of biological diversity (different mammals in Australia and Europe, penguins in the Antarctic but not the Arctic, and so on) had a two-part explanation. God had created all species more or less exactly as they are now; there had been no evolution. Second, he had done this at a large number of discrete centers of Creation. Thus Australia was one such center, South America another. Each had its own flora and fauna. In his botanical textbook of 1836, Henslow, Darwin's friend and former teacher at Cambridge, explained the causes of the geographical distribution of plants in terms of Special Creations followed by "general dispersion from one original spot on the earth's surface." He concluded: "We find the great majority of species so far restricted in their range, as to lead us to the very probable supposition that each was originally assigned by the Creator to some definite spot upon the surface of the earth. . . . From those original creations plants spread widely until interrupted by natural barriers such as seas, rivers, mountains, etc. So far at least forty-five such centers of creation had been identified."[20]

The Special Creation theory was promoted energetically by Louis Agassiz, the world's greatest living naturalist after Alexander von Humboldt. But none of that took into account the existence and diversity of fossils. Nor did it explain extinction: why did God replace one set of species at any one place with a new assemblage? And there was the problem of horses. Darwin found horse fossils in South America, but why was the horse then extinct there? The environment of the New World was obviously well suited for horses, as was demonstrated by their rapid recolonization of north and south after being introduced by the Spanish.

The theory of Special Creation also begged the question of why there were so many species, where one or two might have sufficed. For example, the European buzzard and

American red-tailed hawk look much the same and serve exactly the same role in nature. And, for that matter, why was only one Creator posited, when there could so easily have been many! Perhaps the most logical objection was that, when the fossil record was taken into account, the Special Creation theory implied the occurrence of billions of separate cases of miraculous spontaneous generation—not one of which had ever been observed (the same objection was made to Darwinian transmutation of species, of course).[21]

As he traveled around the world, Darwin had marveled at the fact that each region and often subregion (particularly the islands) had its own fauna and flora. At the Galapagos Islands, moreover, each island—mostly in view of each other—had its own fauna and flora. At this stage, Darwin assumed that all these island birds, tortoises, and plants were all simple variants of species from the mainland, so the level of diversity would not have been remarkable in itself. When the voyage reached Australia, he still held to the Special Creation explanation.

Modern opinion is that it was not until he returned to England and John Gould studied the collections of Galapagos birds and the two kinds of South American rheas, concluding that they were all separate species, that Darwin started to work actively to create a theory of transmutation. His whole five years of being a naturalist—all of his observations and experience, the voluminous reading of his twenty-six years, and his constant questioning and thinking—coalesced around the conclusion that Creation Theory must be wrong. Species must arise from transmutation one from another. The great patterns of nature—the fact that all organisms can be classified into groups on the basis of bodily similarity, the wood-peckers forming one group distinct from the hummingbirds and both different from the finches, for example—all the patterns of similarity and differences and geographical distribution must be due to relationship and to genealogy—to relationship by descent.

Once he had decided that the transmutation of species must be a reality, Darwin soon realized the scope of the task ahead of him and also the enormity of the opportunity. He would have to run straight into the teeth of a popular religious viewpoint, but he was not cowed. To the contrary, he did not simply set out to explain the phenomenon of transmutation of species and discover the mechanism by which it occurred. His goal was far more expansive. He still did not have a mechanism, but he saw where a chance for glory lay. As he wrote in his notebook in 1838, he had in mind a "New System of Natural History" and *"the Grand Question, which every naturalist ought to have before him . . . is 'What are the laws of Life'"* (emphasis in the original).[22]

When it came actually to writing *On the Origin*, his ambitions were clearly set out:

> When the views entertained in this volume on the origin of species, or when analogous views are generally admitted, we can dimly foresee that there will be a considerable revolution in natural history. . . . A grand and almost untrodden field of inquiry will be opened, on the causes and laws of variation, on correlation of growth, on the effects of use and disuse, on the direct action of external conditions, and so forth. The study of domestic productions will rise immensely in value. A new variety raised by man will be a far more important and interesting subject for study than one more species added to the

infinitude of already recorded species. Our classifications will come to be, as far as they can be so made, genealogies; and will then truly give what may be called the plan of creation . . . ? In the distant future I see open fields for far more important researches. Psychology will be based on a new foundation, that of the necessary acquirement of each mental power and capacity by gradation. Light will be thrown on the origin of man and his history.[23]

In *Descent of Man* (1871), he went on to use his theory to explain behavior, morals, and even religion.

I suggested earlier that we are all mixtures of what we know and that conflict is usually engendered when new knowledge or ideas run headlong into the old. This problem of "Ancient and Modern" was particularly acute for Darwin because he was toying with something so completely radical. In developing his theory, Darwin had to examine, reject, and/ or modify a whole panoply of previous scientific ideas (some very ancient, some quite modern) about nature, the material world, and—yes—God. He had to find all the corroborating evidence for his theory that might already exist, conduct his own experiments and observations, and systematically investigate all the rival theories that might impinge on his.

Darwin had already derived a great deal of inspiration, and in fact borrowed the analogy of artificial selection by plant and animal breeders, from the work of his grandfather Erasmus Darwin. He had also delved into the ideas of Buffon, Lamarck's monadism and his inheritance of acquired characteristics, the concept of competition in ecology, a whole range of evidence from embryology, conjectures on the age of earth, extinction, gradual change versus saltation, the whole of natural theology, and MacLeay's Quinarism. Each had to be stripped down to its essentials and, once laid bare, ruthlessly examined for anything useful and everything false.

Dominating all else—the elephant in the room—were natural theology and the hybrid offspring of science and religion forming the "centers of Creation" theory. For Darwin, all theories involving the Creator were essentially "old think." But his "new system of nature" was a dangerous venture. How, then, should he deal with the Creator? Should he allow him to keep his daily starring role or relegate him to the status of first cause only and assign to him only a minor role, if any, thereafter?

In its final form, Darwin's theory was simple and logical. Animals and plants vary. Plant and animal breeders had long since shown that many of the variations are heritable. Those same breeders showed that their intense "artificial selection" could produce striking changes. (One of the most surprising is that cabbage and broccoli are one species.) All animals and plants produce many more offspring than are technically needed to replace them. There is huge wastage and loss in nature; a female cod may lay more than a million eggs in a season. As Benjamin Franklin and Thomas Malthus showed, even humans have the capacity to double their population size in twenty-five years unless checked. This oversupply produces a struggle for existence in an already hostile world. The result is selection of "more favorable" races. The only other thing that is needed to account for the whole panoply of evolutionary diversity is the application of this mechanism of "natural selection" over immense time.

The responses to Darwin's *On the Origin of Species* were many and varied, and in the following chapters I will discuss some of the famous debates that raged during the next few years in Britain and America, the flames subsequently being fanned by the publication in 1871 of his *The Descent of Man*. One of the more delightful (and distinctly negative) responses came from his old geology teacher at Cambridge, Adam Sedgwick:

> My dear Darwin: I write to thank you for your work on the origin of Species. . . . If I did not think you a good tempered & truth loving man I should not tell you that (in spite of the great knowledge; store of facts; capital views of the corelations of the various parts of organic nature; admirable hints about the diffusions, thro' wide regions, of nearly related organic beings; &c &c) I have read your book with more pain than pleasure. Parts of it I admired greatly; parts I laughed at till my sides were almost sore; other parts I read with absolute sorrow; because I think them utterly false & grievously mischievous. . . . I greatly dislike the concluding chapter—not as a summary—for in that light it appears good—but I dislike it from the tone of triumphant confidence in which you appeal to the rising generation & prophesy of things not yet in the womb of time; nor, (if we are to trust the accumulated experience of human sense & the inferences of its logic) ever likely to be found any where but in the fertile womb of man's imagination.[24]

A fascinating attack on Darwin was made by the Reverend Edward Pusey, professor of Hebrew at Oxford, in a famous sermon at the University Church of St. Mary in Oxford in 1878. This was the same Pusey who had earlier given Buckland a philological lifeline for his *Bridgewater Treatises* volume by advising that the words in Genesis usually translated as "created" and "made" were entirely consistent with the Gap

Theory. Ironically, Pusey had also been a major force behind the building of the new Oxford University Museum that, as the locus of the famous (or notorious) Wilberforce-Huxley debate, lost its symbolism as a secular monument to natural theology and became emblematic of modern science.

Now, toward the end of his life, the elderly Pusey was willing to grant science almost all its discoveries. He did not care if the earth turned out to be millions of years old. But he insisted on two points—the divine and miraculous initial creation of life and the exceptionalism of human origins. Other species might evolve, but man required divine intervention. It was a remarkable statement of the way in which the church defended against all those seeds of doubt that had been planted, from Darwin's own work back to the philosophy of the pre-Christian era. It was the new and the old: compromise. "It is not that the book of God's works contradicts the book of God's word, or even that man's interpretation of the one book contradicts his interpretation of the other. They move in two different spheres, and cross each other's path only in the most elementary points. . . . The basis of lasting peace and alliance between physical science and Theology is, that neither should intrude into the province of the other. This is also true science. For science is *certain* knowledge based on *certain* facts. The facts on which Theology rests are spiritual facts; those of physical science are material."[25]

Pusey said that theology can agree with any scientific fact as long as a basic premise is assumed.

> Of the formation of the earth Theology would equally admit of Lucretius' combination of atoms floating in space, and

drawn together by mutual attraction, provided only that He who gave them those impulses, and placed each individual at the distances, whence that attraction would act, was—not chance but God. To Theology creation is equally magnificent, whether the earth started into existence at once at the command of God . . . or whether God created matter, in countless molecules, to be attracted together through a property which He imparted to them, each and all. Theology looks with equal impartiality on all geological theories, atomism, plutonism, neptunism, convulsionism, quietism, provided that, in whatever way it pleased our Creator to act, this be laid at the foundation, that the earth was not eternal, but was created, and that it was at His will, that is passed through whatever transformations it underwent, in conformity to the laws which He imposed upon it.[26]

"Yet," he added, "I must say that, till the properties of Carbon Hydrogen Oxygen and Nitrogen can be made so clear to me, that I cannot conceive of how a soul can result from the sum of them."[27]

So far, so conventional, but then Pusey accused Darwin directly of having written a quasi-theological book, not science. In his sermon (actually, it was in the published endnotes), he said: "It was then, so far, with a quasi-Theological, not a scientific object, that he wrote his book. He wishes 'to overthrow the dogma of special creations.'"[28]

Darwin reacted instantly but, as was usually the case, not publicly, sending off a young admirer, the botanist and geologist Henry Nicholas Ridley, on the errand of rebuttal.

I just skimmed through Dr Pusey's sermon as published in the Guardian, but it did not seem to me worthy of any attention. As I have never answered criticisms excepting those made by scientific men I am not willing that this letter should be published; but I have no objection to your saying that . . . Dr Pusey was mistaken in imagining that I wrote the Origin with any relation whatever to Theology. I should have thought that this would have been evident to anyone who has taken the trouble to read the book, more especially as in the opening lines of the Introduction I specify how the subject arose in my mind . . . but I may add that many years ago when I was collecting facts for the Origin, my belief in what is called a personal God was as firm as that of Dr Pusey himself, & as to the eternity of matter I have never troubled myself about such insoluble questions. —[29]

Darwin's response was quite cleverly calculated and disingenuous. He justified his position on the basis of what he had written in the first page of *On the Origin of Species:*

When on board H.M.S. "Beagle," as naturalist, I was much struck with certain facts in the distribution of the inhabitants of South America, and in the geological relations of the present to the past inhabitants of that continent. These facts seemed to me to throw some light on the origin of species — that mystery of mysteries, as it has been called by one of our greatest philosophers. On my return home, it occurred to me, in 1837, that something might perhaps be made out on this question by patiently accumulating and reflecting on all sorts of facts which could possibly have any bearing on it. After five years' work I allowed myself to speculate on the subject, and drew up some short notes; these I enlarged in 1844 into a sketch of the conclusions, which then seemed to me probable: from that period to the present day I have steadily pursued the same object.

Seeing an opportunity to "throw some light on the origin of species — that mystery of mysteries" was hardly possible for Darwin without considering existing theories and showing that his own was superior. And that he had done with gusto and in any case in 1871 in *The Descent of Man*, to which Pusey must surely have been reacting also, Darwin had made it clear what his aims had been in *On the Origin:* "if I have erred in giving to natural selection great power, which I am very far from admitting, or in having exaggerated its power, which is in itself probable, I have at least, as I hope, done good service in aiding to overthrow the dogma of separate creations."[30]

In his letter to Ridley, Darwin continued in his own defense with what might be optimism, bravado, or arrogance. "Dr Pusey's attack will be as powerless to retard by a day the belief in evolution as were the virulent attacks made by divines fifty years ago against Geology, & the still older ones of the Catholic church against Galileo, for the public is wise enough always to follow scientific men when they agree on any subject; & now there is almost complete unanimity amongst Biologists about Evolution, tho' there is still considerable difference as to the means, such as how far natural selection has acted & how far external conditions, or whether there exists some mysterious innate tendency to perfectibility."

But Pusey had raised a fine point that we can turn around: was the conventional theory of Special Creations a scientific theory or a theological one? Evidently it was both. Darwin's book was from beginning to end precisely what Pusey accused it of being: a demolition of the theory of Special Creations. In that respect, it was indeed a kind of theology, a quasi-theology, as well as a scientific thesis. And it had been so right from the moment he started planning it. But Darwin had not attacked the theory on theological grounds; he viewed his work, therefore, as being entirely scientific. Could a theological proposition be negated by a scientific one and vice versa?

And what of the Creator? As early as 1842 Darwin had been sufficiently sure of his ground and, as usual, worried about his mortality, that he wrote out a short outline of the principle of natural selection, usually termed the *Sketch*. In it, all in all, there are twelve uses of the word *Creator* and four of the word *Creationist*. For example,

> I may premise, that according to the view ordinarily received, the myriads of organisms peopling this world have been created by so many distinct acts of creation. As we know nothing of the [illegible] will of a Creator,—we can see no reason why there should exist any relation between the organisms thus created; or again, they might be created according to any scheme. But it would be marvellous if this scheme should be the same as would result from the descent of groups of organisms from [certain] the same parents, according to the circumstances, just attempted to be developed.
>
> With equal probability did old cosmogonists say fossils were created, as we now see them, with a false resemblance to living beings; what would the Astronomer say to the doctrine that the planets moved [not] according to the law of gravitation, but from the Creator having willed each separate planet to move in its particular orbit? I believe such a proposition (if we remove all prejudices) would be as legitimate as to admit that certain groups of living and extinct organisms, in their distribution, in their structure and in their relations one to another and to external conditions, agreed with the theory.

And then to his conclusions, which are quite recognizable: "Doubtless it at first transcends our humble powers, to conceive laws capable of creating individual organisms, each characterised by the most exquisite workmanship and widely-extended adaptations. It accords better with [our modesty and] the lowness of our faculties to suppose each must require the fiat of a creator, but in the same proportion the existence of such laws should exalt our notion of the power of the omniscient Creator. There is a simple grandeur in the view of life."[31]

In the more completely fleshed out *Essay* of 1844, there are thirteen uses of *Creator* and four references to Creationists. Most of these were recycled one way or another into *On the Origin*, sometimes with "Nature" replacing "Creator." In both the *Sketch* and the *Essay*, the notion that the Creator is the source of diversity and adaptation is always and only presented as a hypothesis to be refuted.

Darwin did not eschew just a touch of sarcasm. "Now these several facts, though evidently all more or less connected together, must by the Creationist (though the geologist may explain some of the anomalies) be considered as so many ultimate facts. He can only say, that it so pleased the Creator that the organic beings of the plains, deserts, mountains, tropical and temperature forests, of S. America, should all have some affinity together; that the inhabitants of the Galapagos Archipelago should be related to those of Chile; and that some of the species on the similarly constituted islands of this archipelago, though most closely related, should be distinct."[32]

Fifteen years later, in the first edition of *On the Origin of Species* itself, there are some eight direct references to "the Creator." All present the views of "Creationists" as the unthinking alternative to a view of evolution through the action of laws and natural selection. There are forty-four uses of the word *Creation*, again all negative. For example, "Natural selection can act only by the preservation and accumulation of infinitesimally small inherited modifications, each profitable to the preserved being; and as modern geology has almost banished such views as the excavation of a great valley by a single diluvial wave, so will natural selection, if it be a true principle, banish the belief of the continued creation of new organic beings, or of any great and sudden modification in their structure."[33]

The very first reference to the Creator in *On the Origin of Species* sets up the debate:

> He who believes in separate and innumerable acts of creation will say, that in these cases it has pleased the Creator to cause a being of one type to take the place of one of another type; but this seems to me only restating the fact in dignified language. He who believes in the struggle for existence and in the principle of natural selection, will acknowledge that every organic being is constantly endeavouring to increase in numbers; and that if any one being vary ever so little, either in habits or structure, and thus gain an advantage over some other inhabitant of the country, it will seize on the place of that inhabitant, however different it may be from its own place.[34]

And, as he put it in the 1844 *Essay*, "Many naturalists have said that the natural system reveals the plan of the Creator: but without it be specified whether order in time or place, or what else is meant by the plan of the Creator, such expressions appear to me to leave the question exactly where it was."[35]

CHAPTER SIX

The Devil and Mr. Darwin

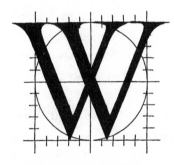

HEN seen from the theological viewpoint of someone like the Reverend Pusey, Darwin's new transmutation theory had several major problematic attributes. It denied a role for the Creator in the diversification of life; it denied the established theory of centers of Creation; it both predicted and required an earth much older than six thousand years; and it fully conformed with atheistic theories of the origin and early history of the earth.

Darwin could not, however, make a firm conclusion about beginnings—the origin of matter, the solar system, or the first life. Since *On the Origin* is set out so clearly and carefully as the alternative to a theological thesis, Darwin's reaction to Pusey's sermon is intriguing. Even more interesting is what he did as soon as the first edition was published.[1]

He blinked.

Darwin had known for twenty years that his theory would be extremely unpopular with both the mass of conventionally religious people and with those scholars for whom religion was serious analytical business. So, even in the first edition, he had dabbled with a compromise. On the inside of the front page he put two epigraphs. The first was harmless enough—a quotation from William Whewell's *Bridgewater Treatises* volume on natural theology that appeared to confirm Darwin's rejection of the theory of Special Creations and centers of Creation: "But with regard to the material world, we can at least go so far as this—we can perceive that events are brought about not by insulated interpositions of Divine power, exerted in each particular case, but by the establishment of general laws." The second epigraph, in the form of a quotation from Francis Bacon's *The Advancement of Learning*, conceded more ground. (A nod to Bacon was almost inevitable to establish the bona fides of a serious scientific work; Newton would have been the other possibility.) Bacon advised that it is not enough to study nature, one must also study God: "Let no man out of a weak conceit of sobriety, or an ill-applied moderation, think or maintain, that a man can search too far or be too well studied in the book of God's word, or in the book of God's works; divinity or philosophy; but rather let men endeavour an endless progress or proficience in both."

So far, so good and so far, so conventional. The quote from Bacon would not have caused any raised eyebrows in surprise. No sooner had *On the Origin of Species* come out, however, than Darwin yielded to temptation and went a step further.

The devil had, as it were, whispered into his ear. "With just a few judicious additions," that insidious inner voice (possibly reinforced by well-meaning friends like the Reverend Charles Kingsley) whispered, "you could appeal to, or at least appease, your religious audience. The acceptance of your ideas would be far greater if you added a phrase or two, emphasizing a role for the Almighty." And Darwin acquiesced. Some combination of fear of rejection and hunger for a truly comprehensive compass for his new system of the world drove him, and we can be pretty sure that he did not act out of any piety because he almost immediately bitterly regretted his weakness.

For the second edition of *On the Origin of Species*, published only weeks after the first (January 10, 1860), Darwin added another, far more theologically acceptable epigraph — a quote from Butler's *Analogy of Revealed Religion:* "The only distinct meaning of the word 'natural' is *stated*, *fixed*, or *settled*; since what is natural as much requires and presupposes an intelligent agent to render it so, *i.e.* to effect it continually or at stated times, as what is supernatural or miraculous does to effect it for once." The key words in that passage were, of course, "intelligent agent." This sentence was essentially an endorsement of the basics of natural theology, something that Darwin in fact rejected totally.

Darwin also added a defensive statement to his second edition: "I see no good reason why the views given in this volume should shock the religious feelings of any one. A celebrated author and divine has written to me that 'he has gradually learnt to see that it is just as noble a conception of the Deity to believe that He created a few original forms capable

of self-development into other and needful forms, as to believe that He required a fresh act of creation to supply the
voids caused by the action of His laws.'"[2]

And then he altered that lovely last sentence. The original
version had been: "There is grandeur in this view of life, with
its several powers, *having been originally breathed* into a few
forms or into one; and that, whilst this planet has gone cycling
on according to the fixed law of gravity, from so simple a beginning endless forms most beautiful and most wonderful
have been, and are being, evolved" (emphasis added). Now
it read: "There is grandeur in this view of life, with its several powers, having been originally breathed *by the Creator*
into a few forms or into one; and that, whilst this planet has
gone cycling on according to the fixed law of gravity, from
so simple a beginning endless forms most beautiful and most
wonderful have been, and are being, evolved" (added words
emphasized). The question that had first been left open by the
two words "originally breathed"—was this meant literally or
figuratively?—had now been answered. Literally.

Darwin made the same addition to the closing sentence
of a paragraph on page 484. "Therefore I should infer from
analogy that probably all the organic beings which have ever
lived on this earth have descended from some one primordial
form, into which life was first breathed *by the Creator*" (added
words emphasized).

On the surface this might be all right, but in fact Darwin
did not believe what he had written. He had added to a scientific work wording for which he knew no justification. This
was real quasi-theology. Darwin soon came bitterly to regret
these changes. He wrote to his friend Hooker in 1863 that he

had "long regretted that I truckled to public opinion & used Pentateuchal term of creation, by which I really meant 'appeared' by some wholly unknown process. — It is mere rubbish thinking, at present, of origin of life; one might as well think of origin of matter. — "[3]

Most unusually for him, when *On the Origin of Species* was reviewed anonymously in the journal *Athenaeum* (probably by John Leifchild) in 1860, Darwin actually wrote to the editor in his own defense.

> He who believes that organic beings have been produced during each geological period from dead matter must believe that the first being thus arose. There must have been a time when inorganic elements alone existed on our planet: let any assumptions be made, such as that the reeking atmosphere was charged with carbonic acid, nitrogenized compounds, phosphorus, &c. Now is there a fact, or a shadow of a fact, supporting the belief that these elements, without the presence of any organic compounds, and acted on only by known forces, could produce a living creature? At present it is to us a result absolutely inconceivable. Your reviewer sneers with justice at my use of the "Pentateuchal terms," "of one primordial form into which life was first breathed": in a purely scientific work I ought perhaps not to have used such terms; but they well serve to confess that our ignorance is as profound on the origin of life as on the origin of force or matter.[4]

Darwin's Pentateuchal additions were not trivial. In deferring to society (and no doubt to Emma), he had given both critics and apologists leeway to argue that he was a closet Christian all along or at least had meant his ideas to be compatible with biblical views of Creation. Those readers who only saw the second edition would not have known that anything had changed, and for many "breathed by the Creator" was just the sort of thing one might expect to read and pass over as a more or less empty gesture. Some scientists, like Asa Gray in America, whom we will meet more fully in the following chapter, actually found it agreed perfectly with their views. To many scientists, in fact, even if they followed Darwin, the idea of removing God entirely from the equation or even restricting his hand to the very first life on earth was both at odds with (contemporary views of) divine revelation and scientifically unnecessary.

When it came to writing *The Descent of Man*, which Darwin launched into with almost a zealot's energy, he tried further to retract his "Pentateuchal" references, reinforcing his opposition to Creation theories. "I am aware that the conclusions arrived at in this work will be denounced by some as highly irreligious; but he who denounces them is bound to show why it is more irreligious to explain the origin of man as a distinct species by descent from some lower form, through the laws of variation and natural selection, than to explain the birth of the individual through the laws of ordinary reproduction. The birth both of the species and of the individual are equally parts of that grand sequence of events, which our minds refuse to accept as the result of blind chance. The understanding revolts at such a conclusion, whether or not we are able to believe that every slight variation of structure—the union of each pair in marriage,—the dissemination of each seed,—and other such events, have all been ordained for some special purpose."[5]

For the rest of his life, Darwin's views were pretty well set and they did not change much except for a vacillation between thinking of himself as an out-and-out atheist and an agnostic. He wrote to Hooker in 1870, for example: "Your conclusion that all speculation about preordination is idle waste of time is the only wise one: but how difficult it is not to speculate. My theology is a simple muddle: I cannot look at the Universe as the result of blind chance, yet I can see no evidence of beneficent design, or indeed of design of any kind in the details.—As for each variation that has ever occurred having been preordained for a special end, I can no more believe in it, than that the spot on which each drop of rain falls has been specially ordained.—Spontaneous generations seems almost as great a puzzle as preordination; I cannot persuade myself that such a multiplicity of organisms can have been produced, like crystals (in the laboratory)."[6]

Basically, he held onto doubt and agnosticism. As he wrote to John Fordyce in 1879: "It seems to me absurd to doubt that a man may be an ardent Theist & an evolutionist. . . . What my own views may be is a question of no consequence to any one except myself.—But as you ask, I may state that my judgment often fluctuates. Moreover whether a man deserves to be called a theist depends on the definition of the term: which is much too large a subject for a note. In my most extreme fluctuations I have never been an atheist in the sense of denying the existence of a God.—I think that generally (& more and more so as I grow older) but not always, that an agnostic would be the most correct description of my state of mind."[7]

Darwin was not passive. He continued to try to find rational arguments for the existence of God. One of them concerned cosmology and the eventual extinction of the sun.

"Believing as I do that man in the distant future will be a far more perfect creature than he now is, it is an intolerable thought that he and all other sentient beings are doomed to complete annihilation after such long-continued slow progress. To those who fully admit the immortality of the human soul, the destruction of our world will not appear so dreadful."[8]

Once *On the Origin of Species* had been digested by the religious, scientific, and lay communities, every possible response, from flat-out rejection to adulation, was published in reviews, letters, and the beginning of what would become a flood of books. Darwin watched all this with a very wary eye and carefully tried not simply to appear to be above the fray but actually to avoid disputes over religion. He assumed the role of a wise man uncommitted to any version of zealotry. "I can hardly see how religion & science can be kept as distinct as [Pusey] desires. . . . But I most wholly agree . . . that there can be no reason why the disciples of either school should attack each other with bitterness."[9] And: "Though I am a strong advocate for free thought on all subjects, yet it appears to me (whether rightly or wrongly) that direct arguments against christianity & theism produce hardly any effect on the public; & freedom of thought is best promoted by the gradual illumination of men's minds, which follows from the advance of science. It has, therefore, been always my object to avoid writing on religion, & I have confined myself to science. I may, however, have been unduly biassed by the pain which it would give some members of my family, if I aided in any way direct attacks on religion."[10]

As for natural theology, which Darwin might be thought

to have effectively killed off, it turned out to be a many-headed hydra. With many an "authority," to demolish a myth scientifically and logically does not automatically dictate that there will not be multitudes who will go on believing it anyway. The reason is simple—that old issue of conflict. And so it was with natural selection. Every individual believer who read and approved Darwin's main conclusions was left with an internal battle. If one side or the other was not to win outright, then compromise was needed. Adopting one version or other of natural theology seemed to offer the far easier option, and today we see that as the concept of "intelligent design."

So many doubts remained, and one nagging question never went away: Darwin could not avoid asking himself and, being intellectually honest, asking his readers also whether the whole idea of God and religion might not have an evolutionary origin and character. He must have been aware that he would be treading on some particularly sensitive toes when he wrote in *The Descent of Man* that:

> There is no evidence that man was aboriginally endowed with the ennobling belief in the existence of an Omnipotent God. On the contrary there is ample evidence, derived not from hasty travellers, but from men who have long resided with savages, that numerous races have existed and still exist, who have no idea of one or more gods, and who have no words in their languages to express such an idea. The question is of course wholly distinct from that higher one, whether there exists a Creator and Ruler of the universe; and this has been answered in the affirmative by the highest intellects that have ever lived.
>
> If, however, we include under the term "religion" the belief in unseen or spiritual agencies, the case is wholly different; for this belief seems to be almost universal with the less civil-

ised races. Nor is it difficult to comprehend how it arose. As soon as the important faculties of the imagination, wonder, and curiosity, together with some power of reasoning, had become partially developed, man would naturally have craved to understand what was passing around him, and have vaguely speculated on his own existence.

The moral nature of man has reached the highest standard as yet attained, partly through the advancement of the reasoning powers and consequently of a just public opinion, but especially through the sympathies being rendered more tender and widely diffused through the effects of habit, example, instruction, and reflection. It is not improbable that virtuous tendencies may through long practice be inherited. With the more civilised races, the conviction of the existence of an all-seeing Deity has had a potent influence on the advancement of morality. Ultimately man no longer accepts the praise or blame of his fellows as his chief guide, though few escape this influence, but his habitual convictions controlled by reason afford him the safest rule. His conscience then becomes his supreme judge and monitor. Nevertheless the first foundation or origin of the moral sense lies in the social instincts, including sympathy; and these instincts no doubt were primarily gained, as in the case of the lower animals, through natural selection.[11]

Darwin clearly conducted a lifelong debate with himself over religion, and he died more or less content that he had done his best in that regard, not definitively closing off the possibility of the existence of God but dubious about his role or roles. There is considerable irony or at least ambivalence in the fact that he was buried in Westminster Abbey, although the abbey is not just a church building, it is Britain's pantheon—and, suitably, Darwin lies not too far away from Newton.

Darwin was always concerned to make sure that his personal doubts did not become a matter of public debate. Now it is time to turn from the private doubts of individuals, as calm as Jefferson's and as tortured as Darwin's, to the more public dilemma that results when the "new" intersects with established beliefs and authority. And to discuss how Darwin's theory of natural selection came to be used by others, often for purposes quite outside of the bread-and-butter issues of science itself.

CHAPTER SEVEN

Debates and Academics

Dear Darwin, I have just come in from my last moonlight saunter at Oxford . . . & cannot go to bed without inditing a few lines to you my dear old Darwin. . . . On Saturday . . . [a] paper of a yankee donkey called Draper on "civilization according to the Darwinian hypothesis" or some such title was being read, & it did not mend my temper . . . however hearing that Soapy Sam was to answer I waited to hear the end. The meeting was so large that they had adjourned to the Library which was crammed with between 700 & 1000 people, for all the world was there to hear Sam Oxon—Well Sam Oxon got up & spouted for half an hour with inimitable spirit uglyness & emptyness & unfairness . . . he ridiculed you badly.[1]

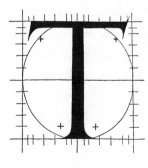HIS is obviously a reference to the famous "Darwin debate" held at Oxford on June 30, 1860, involving, among others, Bishop Samuel Wilberforce and Thomas Henry Huxley. As soon as *On the Origin of Species* had been published, and before Darwin's personal agonies were crys-

tallized in those three words "by the Creator," public and private discussion of his theory had exploded. The Oxford event was the first public debate in Britain of Darwin's theory of natural selection. Unwell as usual, Darwin was not there but left the field to his lieutenants—in the case of the letter just quoted, it was the botanist Joseph Hooker, who clearly thought that it was he who had prevailed against Bishop Wilberforce rather than Huxley, who usually gets whatever credit was deserved.

This meeting in Oxford might in fact be totally forgotten today because little was recorded of what was actually said. It is always portrayed as a contest between bluster and facts. It principally claims a place in the popular history of science because of what Bishop Wilberforce is supposed to have said to Huxley—something very rude about apes and his grandparents. And how Huxley replied—something equally rude about the bishop's intellectual honesty.

The very first public airing of Darwin's theory of evolution by natural selection occurred in July 1858. As is well known, the naturalist Alfred Russell Wallace, while collecting in the Spice Islands in the Indonesian Archipelago, had come up with essentially the same idea of natural selection as Darwin. His twenty-page letter to Darwin, received on June 18, 1858, triggered a joint presentation of the two men's idea at a meeting of the Linnean Society of London on the first of July. The shock finally pushed Darwin into what he had been delaying since at least 1844—writing his book. When *On the Origin of Species* was published in November 1859, all 1,250 copies sold out immediately.

We might assume that the Linnean Society meeting had caused a sensation. Here, surely, was one of those special defining moments when the new comes thunderingly into contact with the old and everyone, protagonists and spectators alike, is forced to take sides. Not a bit of it. It was all an anticlimax. Of the thirty or so Linnean Society members present, some were Darwin's partisans hoping for the best; the others were puzzled, perhaps, but not shocked. The joint idea was neither rejected nor dismissed; it all seemed so abstract. To be sure, it was another interesting idea but not yet one to shake the foundations of natural history, let alone theology. There had been theories on this subject before. Only when *On the Origin of Species*, with all its detailed evidence and argument, had been published did the significance of Darwin's work begin to sink in and come to be discussed widely in scientific and theological circles. Whether or not it was a work of quasi-theology, it was a challenge. Now the hard work of changing minds began in earnest.

We tend to think of the Oxford debate as a pivotal moment, and so it appeared in retrospect. But, oddly enough, by the time of the June 1860 meeting in Oxford, the issue of Darwinism and the origin of new species had already been fought out in the United States, and we have at least partial records of those debates, which were much more substantial than those at Oxford. These debates were conducted at meetings of the American Academy of the Arts and Sciences, the Boston Society of Natural History, and the Cambridge Scientific Club. The protagonists were Asa Gray, professor of botany at Harvard University, William Barton Rogers, geologist and one of the founding professors of the Massachusetts Institute of Technology, and Jean Louis Rodolphe Agassiz, professor of zoology at Harvard.

Gray and Rogers had one thing in common. They hated Agassiz, resenting his assumption of ownership of all American natural science and the self-promoting manner that was working so well to make him a national figure able to fund and create his great legacy, the Museum of Comparative Zoology at Harvard. Anything that Agassiz favored was likely to be opposed by Gray and Rogers. Gray even disapproved of Agassiz allowing his wife to establish a school for girls in Cambridge.

A dominating personality at Harvard and beyond, the Swiss-born Agassiz had made himself an outsize figure in world science. In many ways he was the heir to Humboldt's legacy as the world's greatest naturalist—a role that Darwin was rapidly usurping. Agassiz's early treatise on the fossil fishes of the ancient Old Red Sandstone of Europe (Devonian) was already a classic. Then, in 1837, while still a young man, Agassiz had set the geological and theological universe upside down by conclusively showing that a great deal of Pleistocene geology and all the apparent evidence of a great Flood in Europe and America—valleys carved out, terminal moraines of sand and gravel deposited—was indeed caused by water, but frozen water: glaciers.

Agassiz was an ardent Creationist and committed to his own version of a distinctly continental form of zoological philosophy in the great tradition of Cuvier in Paris. He set out the theory, really a principle, that the whole sweep of nature, living and fossil, manifested three parallelisms: living creatures can be classified on a scale of simple to complex; life on earth has progressed in time from simple to complex; and the embryonic development of individual organisms follows this same progression. It would be said many times during the years after Darwin's theory was established that these phenomena were much stronger evidence for an evolutionary view of life on earth than for one of static Special Creation, but Agassiz saw his threefold parallelism in different terms. As he wrote in his *Essay on Classification:* "The combination in time and space of all these thoughtful conceptions exhibits not only thought, it shows also premeditation, power, wisdom, greatness, prescience, omniscience, providence. In one word, all these facts in their natural connection proclaim aloud the One God, whom man may know, adore, and love; and Natural History must in good time become the analysis of the thoughts of the Creator of the Universe."[2]

A Unitarian, Agassiz outwardly eschewed religiosity, but of all adherents to the idea of Special Creations, he espoused the most extreme, conservative theistic version. Whereas the common European view was that Creation occurred at discrete centers, within which each unique flora and fauna spread over time according to simple geographical circumstances, for Agassiz, such range extension in time did not occur; each species had a given range at Creation that never changed. For him, the present species were originally created in about the same proportionate numbers as they are found to have at the present time, exactly as they are now, and in precisely the same localities as those they now occupy. This in turn required an even more vast number of spontaneous generations than conventional Special Creation Theory required.

And Agassiz extended his views to humans. The question whether human beings constitute a single variable species or a number of distinct species had been debated heatedly since the mid-eighteenth century. Earlier in his career Agassiz had believed that all humans were one species. Then in America he had direct experience of a wider range of human beings than he ever knew in Neufchatel, Switzerland, and he changed his mind. A serious obstacle for him was that the evident variation that occurs in and among human races (he distinguished eight: Caucasian, Artic, Mongol, American Indian, Negro, Hottentot, Malay, and Australian) ran fully counter to his view of Special Creation, in which he allowed no variation within any species at all. Further, if the races of humans had diverged from some single type (Caucasian, of course), then that would have conformed all too well with a "developmental" evolutionary view such as Lamarck's. Thus his student Edwin Sylvester Morse, later an ardent Darwinian, wrote in his diary for April 26, 1860: "Splendid lecture by Prof this morning on the absurdity of believing that Adam and Eve were the first created and the only ones. It was a masterly lecture and was listened to with great attention."[3]

Instead, Agassiz believed that human races had been created as separate species in situ—Europeans in Europe, Africans in Africa, Australian aboriginals only in Australia, and so on. This, unsurprisingly, was a point of view that gave comfort and support to southern slave owners.

It was a given of all versions of the Special Creation thesis that the same species could not occur in two different faunas or floras. There could in principle be no species in common between, say, Europe and Africa, or Asia and North America, except through human intervention. Finally, Agassiz contended that no species currently living had existed before Pleistocene times. Like his hero Cuvier, he believed that all fossil species belonged to previous rounds of creation and extinction, requiring yet more waves of spontaneous generation.

Asa Gray, America's leading botanist, was the immediate link between Kent and Massachusetts. Like Charles Darwin, he marveled at both the diversity of different plants on the surface of the earth and at the patterns of their distribution. Just like Darwin, he was challenged to explain why there were so many species and what caused the obvious patterns of similarity (relationship at least of structure) among them. Gray had corresponded with Darwin since 1855, which was just the time when Darwin started privately to share his idea of natural selection as a mechanism for evolution in which the patterns of diversity in space and time represented a genealogy of life. He had begun seeking both advice and approval within a small cadre of close, trusted colleagues and had thereby also started in a quite political sense to build up a body of positive support within the scientific fraternity, so that when he finally would reveal his ideas to the world, there would be a cohort of enthusiasts ready to receive them favorably and offset the expected naysayers.

Darwin had gradually let Gray into his evolutionary secret and sent him a draft of his theory in September 1857. That turned out to be useful when Wallace dropped his bombshell on the bucolic peace and intellectual ferment of Down House, Kent. Darwin was able to use his correspondence with Gray to demonstrate his intellectual priority. Gray was surely the first American to know of Darwin's theory. It is not clear just when the first copies of the *Proceedings of the Linnean Society* with the Darwin-Wallace papers arrived in America—but it was probably no later than the fall of 1858. Copies of *On the Origin of Species* arrived in December 1859. By then the debates had already begun.

Beginning in 1840, Gray had been interested in comparisons of the flora of Japan and North America, a study that was facilitated when he was given access to a large collection of plants made by the Ringgold-Rogers North Pacific Exploring and Surveying Expedition of 1853 to 1859. In January 1859 he submitted the results of his work for publication by the American Academy of Arts and Sciences.[4]

In his long *Memoir*, Gray developed startling empirical evidence that the theory of centers of Special Creation was wrong. Instead of the floras of Asia and North American being totally distinct, as theory required, Gray showed the opposite. With the great advantage of knowing that in his pocket, as it were, was work by Darwin that would authenticate his ideas and set them within a new worldview of nature, Gray set out a case study in evolutionary biogeography and, it is not too much of an exaggeration to say, launched a campaign against Agassiz.

The first unveiling of Gray's work came in the form of a presentation delivered to the December 10, 1858, meeting of the Cambridge Scientific Club, an informal dining club mostly of Harvard professors. This was only six months after the Linnean Society meeting in London. Agassiz was said to have remained strangely silent after the presentation. On December 18, 1858, Gray made an open attack on Agassiz when giving a presentation of his work to the American Academy of Arts and Sciences, continuing at the meeting of January 11, 1859, and several further meetings beyond. This was almost a year before Darwin's book would come out and eighteen months before the Oxford debate. And here was an important difference. At Oxford science was on the defensive; here the new evolutionary science went on attack.

Gray dwelt at length on the fact that "the relations of the Flora of Japan with those of the United States east of the Mississippi were peculiarly intimate, as evinced by the great number of congener, of closely representative, and of identical species . . . absent from the flora of Europe." There could be no doubting that this meant there had been "a remarkable interchange between the floras of Eastern North America and Eastern Asia . . . a former homogeneousness of the temperate American and east Asian floras, to a degree equal, perhaps, to that of the Arctic or the sub-Arctic flora at the present time."[5]

This odd-seeming connection, applying mostly between the southeastern United States and eastern Asia, with many more groups in common than between eastern Asia and northwestern America, is now well documented. Examples include the familiar camellias. In Gray's time this discovery was more than disconcerting; it was exactly what Agassiz denied. Gray explained the patterns in terms of major long-term shifts in climate affecting the distribution of floras. "This interchange had been facilitated or driven by significant shifts in climate such that the region of the Bering Straits had once been far more equable and species could cross over." If that were not the explanation, the two "centers"—Asia and North America—had not been distinct at Creation. It was a complicated model: "The temperate flora of North America and of Western Asia—before coterminous, and then most widely separated—must have again become coterminous . . . a milder climate than the present then supervened Mastodon etc ranged north and rhinoceros in the Old World, across the Bering Straits. Further, since it is conceded that the present era of the world is of extremely long duration, and since it is most probable, not to say certain, that the existing species of plants in the regions in question, or a part of them, are of high antiquity, dating back to the post-tertiary, or even to the later tertiary epoch, and therefore must have been subject to great climatic changes, accompanied or caused by no inconsiderable changes in the relative extent and configuration of the land."[6]

In many respects, Gray's summary of his argument, as recorded by the academy secretary, conformed perfectly with Darwin's thesis in *On the Origin of Species*. Gray's model was a dynamic one of origin and dispersion. He argued in Darwinian terms that

> the idea of the descent of all similar or conspecific individuals from a common stock is so natural and so inevitably suggested by common observation, that it must needs to be first tried upon the problem. The creation of beings endowed with such enormous multiplying power, and such means and faculties for dissemination, as most plants and animals. Why then should we suppose the Creator to do that supernaturally which would be naturally effected by the very instrumentalities which he has set in operation? And in his [Gray's] opinion the actual question now is,—whether each species originated in one local

area, whence it has spread, as circumstances permitted, over more or less broad tracts, in some cases becoming discontinuous in area through changes in climate or other physical conditions operating over a long period of time; or, whether each species originated where it now occurs, probably in as great a number of individuals occupying as large an area, and generally the same area, or even the same discontinuous areas, as at the present time. The latter is understood to be the view of Professor Agassiz.

It was a classic contrast, and Gray continued that Agassiz's theory "offered no scientific explanation of the present distribution of species over the globe; but simply supersedes explanation, by affirming, that as things now are, so they were at the beginning; whereas the facts of the case—often very peculiar—appear to demand from science something more than a direct reference to the phenomena as they are to the Divine will."[7]

As Morse's biographer puts it, "Gray himself, writing a friend after one of these debates, said gleefully, 'I knocked out the underpinning from Agassiz' theories about species annent local creation of species, turning some of Agassiz' own guns against him.' It was bear-baiting, in short, with Agassiz tortured far beyond his friends's understanding and the young students watching as gleefully as though it were a pit with a bear beleaguered by a band of bulldogs."[8]

Meetings of the American Academy were conducted with somewhat greater decorum than the rowdy Oxford meeting; most of the discussants were Harvard colleagues. So Gray had gone out of his way to acknowledge the opposing explanation—the Special Creation Theory—knowing full well that the person who, in the whole world of science, held the most extreme views on Special Creation was none other than Louis Agassiz.

Rising to reply to Gray, Agassiz acknowledged that his views had been accurately represented and he replied that the pattern seen by Gray extended to the animals also. But when Gray knew his subject better, his theory would fall apart from lack of evidence. There is not at the present time, he added, "an equal knowledge of all the facts in Botany and Zoology."[9]

Agassiz explained the similarities between the two regions "as a primitive adaptation of organic types to similar corresponding physical features, which have remained respectively unchanged since the first introduction on earth of these organisms." In other words, things looked similar but were not. Further, "the present period might be immensely long, but physical changes could not have produced the results Gray claimed. And, as for the single origin of conspecific individuals, he thought that the warfare which some say species wage upon others was itself an insuperable objection to the assumption that any one species could have originated in a single pair."[10]

Everything in Gray's interpretation of the distribution of the key Japanese and North American species was explicitly denied by Agassiz. Countering, Gray went out of his way to state that it was illogical to believe that the same plant species had been separately created in different parts of the world. All Agassiz could say was that "when we speculate about the origin of species, we launch out beyond the region of induction, and have only analogues or probabilities to guide us."

The subject was taken up at a subsequent supplementary

meeting. On February 22, 1859, Gray opened a new line of attack, now examining the fossil record. Agassiz had claimed that no modern species had lived in the pre-Quaternary periods. Gray quoted Agassiz's close friend (and fellow Swiss) the paleobotanist Leo Lesquereux as having found live oak, honey locust, pecan elm, *Ceanothus*, *Prunus caroliniana*, and *Quercus myrtifolia* in the Lower or Middle Pliocene. These are all species now living on the coast and islands of the southern states. Agassiz countered that this again was a result of inexpert identification.[11]

Copies of Darwin's *On the Origin of Species* reached America around Christmas 1859, almost a year after the first debate at the American Academy. Gray wrote instantly to Darwin and to his botanical colleague Hooker, making it clear that he was applauding not only a good theory but the downfall of Agassiz. To Hooker he wrote, "Well, the book has reached me, and I finished its careful perusal 4 days ago!—And I freely say that your laudation is not out of place. It is done in a *masterly manner*,—it might well have taken 20 years to produce it. It is crammed full of most interesting matter—thoroughly digested—well expressed—close, cogent—and taken as a system it makes out a better case than I had supposed possible. . . . Agassiz—when I saw him last, had read but a part of it. He says it is *poor—very poor!!* (entre nous). The fact he growls over it, like a well cudgelled dog,—is very much annoyed by it—to our great delight—and I do not wonder at it."[12]

In March 1860 Gray published a long review of *On the Origin of Species* in the *American Journal of Science*, America's leading scientific journal, and over the next months he wrote several other reviews, including a long two-part essay for a general audience in *Atlantic Monthly*. Agassiz answered with his own review of Darwin in the *American Journal of Science*.

Gray helped get Darwin's book published in the United States, and now the debates started again. The Creationists fought back. At the American Academy's April 10, 1860, monthly meeting, Professor John Amory Lowell gave a comprehensive critique of Darwin's book, proclaiming Darwin's failure to demonstrate the reality or extent of variation in nature and to demonstrate the validity of an analogy between artificial and natural selection. "Man acts with means of seclusion [selection] that Nature does not," he stated, and "Changes produced by human agency are all within specific limits. . . . The most improved South-down ram, or Ayrshire bull, is but a ram or a bull after all. You cannot, therefore reason from this analogy, whatever time be assumed, to any changes differing in *kind*." He worried over the notion that variations were "accidental" and argued that the fossil record does not support "an ascending series, while the lower forms were extinguished," pointing to the living fossil (brachiopod) "clam" Lingula as an example.[13] Professor Francis Bowen (another friend of Agassiz from the Harvard faculty) raised the long-standing objection that Darwin's hypothesis required too much time. In fact, it made "such huge demands upon time, that the indefinite becomes virtually infinite time, so rendering the theory dependent on metaphysical rather than inductive reasoning."

The whole thing was fought over again at the May 1, 1860, special meeting. Each time Gray was at pains to point out (what would become a familiar defense) that he had never presented Darwinian theory "as anything more than a legitimate hypothesis, just beginning to stand its trial." He maintained that "the varieties of cultivation afforded direct evidence of the essential variability of species . . . no domesticated plant had refused to vary. . . . Man produces no organic variation, but merely directs a power which he did not originate, and by selection and close breeding preserves the incipient variety . . . and gives it a choice opportunity to vary more." As to Professor Bowen's complaint that the (almost) infinite time that the theory required rendered it "completely metaphysical in character," Gray remarked that in fact the theory "would generally be regarded as too materialistic and physical, rather than too metaphorical; and that, *a fortiori*, physical geology and physical astronomy would on this principle [also] be metaphysical sciences."[14]

The trickier criticisms to answer were the theological ones. Bowen had charged that Darwin's theory was incompatible with final cause (purpose) and with the "argument for design" (natural theology). Gray answered that regardless of how a change in species was brought about, it was "all the same as to the argument for design, this resting on the adaptation of structure to use, irrespective of the particular manner in which the adaptation may be conceived to have been brought about." Natural selection is an "efficient cause," and Gray argued that either (as Bowen said) "the origination of an individual, no less than that of a species, requires and presupposes Divine power as its efficient cause" or "the origination of a species is natural, no less than the origination of an individual; — propositions which do not appear to contradict each other."[15]

If all this were not enough, the cudgels had also been taken up in meetings of the Boston Society of Natural History. Here, Agassiz went on the attack himself, again loftily claiming that what others saw as evidence was due to their inexperience as systematists. He stood by his position that all species were created invariant. If there were no variation in nature, then a key part of Darwin's theory failed. But this meant that Agassiz had to explain away what others saw as variation, which he did simply by saying they were mistaken. They were really seeing different species, as Gray would discover, he loftily pronounced, when he was more experienced.

Agassiz now developed a new line with respect to the fossil record. He seized on living fossils (the concept was Darwin's) to show that evolution did not happen and repeated the claim that species thought to have been living before the Pleistocene were misidentified and that the fossil record proved that no species living today was extant in the Tertiary. (Of course, this was self-contradictory: the case of the living fossils proved that modern species had existed in ancient beds.) This time William Barton Rogers took up the challenge. It was easy to give examples of living species that also existed as Tertiary fossils. It was not going to be enough for Agassiz to claim that every paleontologist alive was incompetent.

At a subsequent meeting, Darwin's geology came up again. Agassiz had the opinion that the Paleozoic beds of North America had been laid down in periods of upheaval and subsidence of the seabed. Darwin and the two Rogers brothers, together with Charles Lyell, had interpreted that they were laid down on a shallow seabed that slowly subsided. Otherwise, as W. B. Rogers cuttingly pointed out, Agassiz would have had to believe that the seas over eastern North America had once been two or three miles deep. Surprisingly, Agassiz admitted the error.

It was momentous. Cumulatively, these sets of meetings had been a disaster for Agassiz. They had been a raw, if genteelly conducted, contest for authority and particularly a tearing-down of Agassiz's previously uncontested claim to pronounce on all matters of natural history. Agassiz's scientific influence had begun to wane just as his political presence continued to grow. Interestingly, while he continued to argue against evolution in popular reviews and articles, Agassiz never again addressed the subject in a scientific context.

In these debates Agassiz himself had not overtly argued from a theological point of view. He did not appeal to revelation. It was his scientific authority that was on the line. On the quasi-theological versus quasi-scientific spectrum, he treated Special Creation first and foremost as demonstrable science. In fact, religion was largely absent from the Boston debates. Was that because the social atmosphere in Boston was very different from Oxford and London when it came to discussing religion and arguing about religious elements in science in public?

Perhaps one reason is that Gray had a religious card up his sleeve. In his critique of Darwin, J. A. Lowell had charged that "the hypothesis in question repudiates design or purpose in nature and the whole doctrine of final causes." But the subtitle of the book-length version of Gray's reviews of Darwin in the *Atlantic Monthly* was *Natural Selection Not Inconsistent with Natural Theology*. Gray had in mind a scheme far better than any Special Creation Theory. It was all very simple. For Gray, variation was not random in occurrence, as Darwin claimed, or directed by environmental pressure, as Lamarck proposed (a position Darwin had started to edge toward). Instead it was directed by God. This is another superb example of saving the phenomenon or having your cake and eating it too. By adding a divine element to natural selection, Gray neatly replaced the religious authority inherent in Agassiz's Special Creations with a dynamic system. By allowing for the action of the divine hand in variation, Gray removed the problem that so many found difficult to accept—the role of pure chance.

Gray's idea of the hand of God in variation made the whole process of evolution rather cumbersome, however. Why would God do indirectly what the Special Creationists said he could do directly? *"Why . . . should we suppose the Creator to do that supernaturally which would be naturally effected by the very instrumentalities which he has set in operation?"* (emphasis added).

Gray's slightly less than wholehearted acceptance of Darwin's theory showed up even more strongly in 1871 when *The Descent of Man* appeared. He wrote to Darwin after reading the first part of the book, "Almost thou persuadest me to have been a hairy quadruped, of arboreal habits, furnished with a tail and pointed ears." But the important word is "almost." Whereas he had rushed to publish review after review of *On the Origin of Species*, with respect to *Descent of Man* he made an excuse: "I have been besought to write notices of the book. But I decline. You don't know how distracted I am in these days."[16]

In fact, no major American scientist would dare write a positive review of that book for decades to come.

CHAPTER EIGHT

Clerics and Apes

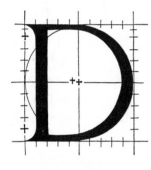 ARWIN'S timing, waiting for the furor to die down over Chambers's 1844 *Vestiges of the Natural History of Creation*—a retooling of Lamarckian transmutation theory—before publishing his own theory, was perfect. Evolution in one guise or another and as an insidiously growing issue in the philosophical and societal background was steadily becoming accepted. Despite (or because of) religious objections, it was a ripe time for a truly scientific mechanism to be discovered. For example, in 1859 the Reverend Baden Powell wrote what was beginning to be obvious: "It is absurd to argue that the introduction of new forms of life, or new species of organic beings, in the successive epochs of the earth's formation, involves a peculiar mysterious power, or supernatural creation, merely because we do not at present know the cause

of life, or see new species arise before our eyes, which, it may be added, we never could detect as such if they did."[1]

It was also Darwin's good fortune that by 1859 geologists had produced a system that largely reduced the biblical account of Creation to a fable and had established the extreme age of the earth that was necessary for his processes of natural selection to work major changes. Even more conveniently for Darwin, the religious establishment was busily engaged in its own bitterly rending debate over the extent to which the new discoveries should be incorporated in religious teaching and orthodoxy. That debate was neatly laid out in the book *Essays and Reviews* (1860), a collection of seven essays by six leading clerics and one layman. Of the many critical issues that were addressed in *Essays and Reviews,* the greatest problems had to do with the Bible and divine revelation when reconsidered in the light of the German literary and philological researches and the Gap Theory. The Reverend Rowland Williams's essay on the German rationalist scholar Christian Bunsen drew the most attention at the time, while today the book is more remembered (at least among those scientists who know anything of it) for an essay by Rev. Powell fully transferring his support to Darwinism (surely the first such approval published by any cleric).

A good moment for Darwin was a bad one for England's established church. And it all set the stage for the annual meeting of the British Association for the Advancement of Science, which had been scheduled for Oxford in June 1860 in large part to inaugurate and celebrate the wonderful new University Museum, which has been called a secular temple to science and nature.[2] Samuel Wilberforce, bishop of Oxford (and son of the more famous William Wilberforce, abolitionist), was chairman of the local committee organizing the meeting, and it must have seemed to him a heaven-sent opportunity to restore some order among both clergy and laity. And while in the Boston debates the proponents of the new evolution were on the offensive, at Oxford Wilberforce intended to put them on the defense.

At the meeting nearly two hundred papers were read, covering a wide variety of subjects, but Darwin's new theory of natural selection seemed to hang over a lot of the discussion. As a contemporary report put it,

> The chief cause of contention has been the new theory of the Development of Species by Natural Selection—a theory open—like the Zoological Gardens (from a particular cage in which it draws so many laughable illustrations)—to a good deal of personal quizzing, without, however, seriously crippling the usefulness of the physiological investigations on which it rests. The Bishop of Oxford came out strongly against a theory which holds it possible that man may be descended from an ape,—in which protest he is sustained by Prof. Owen, Sir Benjamin Brodie, Dr. Daubeny, and the most eminent naturalists assembled at Oxford. But others—conspicuous among these, Prof. Huxley—have expressed their willingness to accept, for themselves, as well as for their friends and enemies, all actual truths, even the last humiliating truth of a pedigree not registered in the Herald's College. The dispute has at least made Oxford uncommonly lively during the week.[3]

(The "pedigree" sentence was obviously a discreet reference to "apes for ancestors.")

As is well known, the morning session of Saturday, June 30, began with a major address by Professor John W. Draper, an Englishman transplanted to New York. He was a distinguished chemist and physiologist who had taken up the history of science, and his subject was the ways in which the histories of nations develop—evolve—from primitive to advanced just as organisms do. It was a sort of ur-social Darwinism, although it might also have been an extension of Agassiz's threefold parallelism. Wilberforce was set up to answer and was universally expected to use the occasion to take on Darwin directly.

Everything about the debate is now upside down. For example, Draper (whose part in the actual meeting tends to be forgotten or dismissed as a bore) eventually played a pivotal role by writing, fifteen years later, *History of the Conflict between Religion and Science*, the first modern exposition of the religion-science controversies. In the mythic account, the scientists turned up that morning anticipating a brutal fight; in fact, they seem not to have been completely prepared for it. Darwin was not there—one might also add "of course." His delicate stomach would never have withstood such a public airing of his ideas. Huxley had not planned to attend either and was persuaded to do so only at the last minute by none other than the discredited Robert Chambers. While the debate was covered in all the newspapers at the time, most contemporaries forgot about it until Huxley revived it thirty years later when he wrote his autobiography. But we are talking about it still.

As to Bishop Wilberforce's address, he was widely assumed to have been coached in his biology by Professor Richard Owen in London, a rival of Darwin. Wilberforce published a long review of Darwin's book that year in which the hand of Owen can clearly be seen, and that review was surely the foundation of Wilberforce's Oxford address. That essay has been used here (in the appendix) as the basis for an attempted reconstruction of what Wilberforce said. In the only more or less authentic contemporary account, the *Athenaeum* reported:

> The BISHOP OF OXFORD stated that the Darwinian theory, when tried by the principles of inductive science, broke down. The facts brought forward did not warrant the theory. The permanence of specific forms was a fact confirmed by all observation. The remains of animals, plants, and man found in those earliest records of the human race—the Egyptian catacombs, all spoke of their identity with existing forms, and of the irresistible tendency of organized beings to assume an unalterable character. The line between man and the lower animals was distinct: there was no tendency on the part of the lower animals to become the self-conscious intelligent being, man; or in man to degenerate and lose the high characteristics of his mind and intelligence. All experiments had failed to show any tendency in one animal to assume the form of the other. In the great case of the pigeons quoted by Mr. Darwin, he admitted that no sooner were these animals set free than they returned to their primitive type. Everywhere sterility attended hybridism, as was seen in the closely-allied forms of the horse and the ass. Mr. Darwin's conclusions were an hypothesis, raised most unphilosophically to the dignity of a causal theory. He was glad to know that the greatest names in science were opposed to this theory, which he believed to be opposed to the interests of science and humanity.[4]

Wilberforce adeptly picked at the weak points of Darwin's theory—which left itself open to attack because it was after all presented as "one long argument" rather than proven. He dissected Darwin's book point by point, brilliantly adding to his impressive logic a withering sarcasm. A final conventionally religious note brought the subject squarely back to his audience, of whom the great part were cheering clerics. But how did Wilberforce end his peroration? He knew that he had carried the audience with him, being interrupted by cheers, jeers, and laughter at Darwin's expense. Did he then, carried away by his own rhetoric, end by singling out Huxley sitting in the audience in front of him and questioning his ancestry?

We do know that Huxley, furious, rose to answer the bishop with a review of the strong points of Darwin's theory, which he admitted was still incomplete. In an answer to the perennially continuing charge that evolution is "just a theory," Huxley said that

> it was an explanation of phenomena in Natural History, as the undulating theory was of the phenomena of light. No one objected to that theory because an undulation of light had never been arrested and measured. Darwin's theory was an explanation of facts; and his book was full of new facts, all bearing on his theory. Without asserting that every part of the theory had been confirmed, he maintained that it was the best explanation of the origin of species which had yet been offered. With regard to the psychological distinction between man and animals; man himself was once a monad—a mere atom, and nobody could say at what moment in the history of his development he became consciously intelligent. The question was not so much one of a transmutation or transition of species, as of the production of forms which became permanent. Thus the short-legged sheep of America were not produced gradually, but originated in the birth of an original parent of the whole stock, which had been kept up by a rigid system of artificial selection.[5]

But Hooker is the one who really fought back. Hooker turned the tables on Wilberforce in part by pointing out that he had been attacking the wrong theory, namely, Lamarck's. The only contemporary report stated:

> In the first place, his Lordship, in his eloquent address, had, as it appeared to him, completely misunderstood Mr. Darwin's hypothesis: his Lordship intimated that this maintained the doctrine of the transmutation of existing species one into another, and had confounded this with that of the successive development of species by variation and natural selection. The first of these doctrines was so wholly opposed to the facts, reasonings, and results of Mr. Darwin's work, that he could not conceive how any one who had read it could make such a mistake, — the whole book, indeed, being a protest against that doctrine. Then, again, with regard to the general phenomena of species, he understood his Lordship to affirm that these did not present characters that should lead careful and philosophical naturalists to favour Mr. Darwin's views. To this assertion Dr. Hooker's experience of the Vegetable Kingdom was diametrically opposed. He considered that at least one half of the known kinds of plants were disposable in groups, of which the species were connected by varying characters common to all in that group, and sensibly differing in some individuals only of each species; so much so that, if each group be likened to a cobweb, and one species be supposed to stand in the center of that web, its varying characters might be compared to the radiating and concentric threads, when the other species would be represented by the points of union of these; in short, that the general characteristics of orders, genera, and species amongst plants differed in degrees only from those of varieties, and afforded the strongest countenance to Mr. Darwin's hypothesis.[6]

When Hooker wrote his letter to Darwin afterward, he was obviously still thrilled by the occasion and his own contribution. We can forgive him some hyperbole, even if we are little surprised at the vehemence of the language. "Sam Oxon

got up & spouted for half an hour with inimitable spirit uglyness & emptyness & unfairness . . . he ridiculed you badly &
Huxley savagely—Huxley answered admirably & turned the
tables, but he could not throw his voice over so large an assembly, nor command the audience; . . . now I saw my advantage—I swore to myself I would smite that Amalekite Sam hip
& thigh . . . & there & then I smashed him amid rounds of
applause. Sam was shut up—had not one word to say in reply
& the meeting *was dissolved forthwith* leaving you master of the
field after 4 hours battle. I have been congratulated & thanked
by the blackest coats & whitest stocks in Oxford."

There is no account of Wilberforce having replied to
Hooker. From the newspaper reports of the day, however, two
things are clear, or fairly clear. At the end of his address Wilberforce *did* turn to Huxley and say something about parentage
and apes. Possibly it had been something like this: "If what
Mr. Darwin has proposed were true, then we should all be
descended from monkeys. I wonder—Professor Huxley—
whether you would be so good as to tell the audience whether
it is on your grandmother's side or your grandfather's that you
claim descent from an ape?" It is more certain that, at the end
of his own remarks, Huxley answered Wilberforce with the
extremely pointed remark that he was not ashamed to have an
ape for an ancestor, but he would be ashamed to be connected
with a man who used great gifts to obscure the truth.

Whatever the exact wording, these exchanges were, especially for Victorian times, the equivalent of medieval knights
slapping a glove across the face. There seems to be no doubt
that Huxley essentially called Wilberforce a liar, but did
Wilberforce really say "grandmother" and why?

The monkey business, so to speak, turns out to have been instigated by Huxley himself on the previous Thursday at a scientific session when he had got into a heated debate with Owen over the similarity of the gorilla and human brains. Owen, a master anatomist who surely knew how similar they were, claimed that they were vastly, incompatibly, different. Huxley knew otherwise. It was at this session that the question of human ancestry from apes was discussed favorably by Huxley as being supported by scientific fact, although obviously it had not occurred at a mere two generations' remove. The "great feature which distinguished man from the monkey" was not anatomy but "the gift of speech." So Wilberforce may actually have said something like, "I gather that Professor Huxley has said two days ago that he would not be ashamed to claim ancestry from the apes. I wonder if he would be so good as to tell us whether it is . . . etc." That would have been a piece of perfect invective and sensational for an audience of clerics, students, and lady visitors.

As usual, however, things are not that simple. The *Glasgow Herald*, which took its reports of the meeting from a local stringer ("an able and local townsman"), reported things rather differently on July 4, 1860: "The Bishop of Oxford, with all the authoritative dogmatism of the episcopate, denounced the idea ['that the progress of organisms is determined by law'] as absurd, although he allowed that, to any able or sensible man, *it was of little consequence to himself* whether or not his grandfather might be called a monkey or not" (emphasis added).

In this second account, Wilberforce's attitude toward ape ancestry is one of lofty, supercilious indifference, quite different from the traditional version in which he made a sarcastic ad hominen attack. In the end we will never know what, in that heated debate, the bishop of Oxford really said. But while it is just possible to accept that Wilberforce had referred specifically to Huxley's grandfather or great-great-grandfather, it does seem unlikely that in those days a bishop could have, effectively, called Huxley a son-of-bitch by even hinting at a gorilla grandmother. Not a bishop of Oxford, however carried away by his own rhetoric. If he did, no wonder Lady Brewster fainted.

These debates or confrontations—whether in Oxford or in Massachusetts—made and still make wonderful theater, perhaps especially because we do not have the full script of what was said. Whether it was Wilberforce or Agassiz, the theme of the drama to modern eyes is one of giants brought low, of authority challenged, and of bluster and dogmatism being overwhelmed by new facts and logic (together with the purely personal side, the animosities and allegiances).

All the debates capture a more-or-less intermediate stage in the complex process of translation of the ideas of individuals into new institutional authority. They can be seen also as two way stations in a long and continuing intellectual journey in which religious authority responded to a complex mixture of new secular nonscientific and scientific ideas and facts.

At first little changed, and yet everything had—if for nothing else, for one issue of profound importance. Darwin theorized, and paleontologists soon showed, that man is not a Special Creation of God's but really is descended from apes. Copernicus removed the earth from the center of the universe: Darwin gave humans a pedigree and transplanted God from being the Special Creator to at best the status of the author of only the initial laws of matter.

In Oxford on the Sunday morning after the debate, Frederick Temple, one of the authors of *Essays and Reviews*, gave the sermon at the University Church of St. Mary. Instead of conflict, he preached compromise and conciliation, developing the old theme that a man of science need not be a nonbeliever. "The student of science now feels himself bound by the interests of truth, and can admit no other obligation. And if he be a religious man, he believes that both books, the book of nature and the book of Revelation, alike come from God, and that he has no more right to refuse to accept what he find in one than what he finds in the other."[7] That continued, and continues, to be a popular default position.

CHAPTER NINE

The Decline of Authority

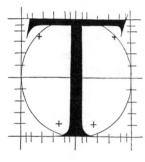HE Oxford and Massachusetts debates were all about change and challenges to authority, both of which are as inevitable as Franklin's "death and taxes." Everything changes. The charge by Mr. Terry to those who would deliver his lectures specifically mentions evolution. I have already discussed evolution in its technical, biological sense, but the word can also be used more generally for change. Most change in life is like biological evolution—contingent. Very little arises completely de novo; everything derives from something else and is both inspired and constrained by what has gone before. And change is everywhere; to pretend that one can resist it is foolishness; wisdom lies in how one deals with it.

Fairness requires us to recognize the fact that authority

and its essential intransigence (or, more politely, conservatism) are hardly unique to the religious sphere. Science has its own sects and its own beliefs. Science has its own texts, its own high priests and shibboleths. Science and scientists can be equally resistant to change.

At school in the 1950s, I was taught that continental drift was proven by Alfred Wegener's maps showing the fitting together of the western and eastern coastlines of the Atlantic. As a college student, I was taught that that was nonsense—how could continents have moved? Then, when I was a graduate student, I was lucky enough to hear a presentation at the Royal Society by Steven Runcorn giving striking evidence of seafloor spreading at the mid-Atlantic Ridge. It was hard evidence and the beginning of a mechanism—plate tectonics—for continental drift. Returning to Harvard that fall, I discovered the professors of geology bitterly divided, many still sticking to the theory of stationary continents.

In another example, in the 1970s the late Sherwood Rowland discovered the danger for the ozone layer created by the release into the atmosphere of chlorofluorocarbons (in refrigerator coolants and a whole range of aerosol applications). His work was discounted (especially by the industrial community) before the evidence became overwhelming. And the way in which the ozone hole over the Antarctic was discovered and Rowland was vindicated presents another object lesson. American flights over the South Pole consistently failed to find the predicted "ozone hole." Then a British program using old planes and more primitive computing discovered it. It turned out that the American computerized analytical programs had been set up to dismiss data that exceeded some preset norm. One wonders whether in other aspects we are so preprogrammed to find only one set of phenomena that we miss others.

Context is important. We cannot discuss the origin of Darwinian evolution without remembering that it rose to the fore alongside the massive industrialization of society, new political, economic, and even legal philosophies, increases in every kind of personal freedom, and startling discoveries about the nature of the physical world and cosmos. And with them came whole new movements in society involving change rather than stasis and even involving revolution against all kinds of authority.

What is not to be questioned, however, is that with each loss—or rather each change—in authority, new doubts arose as well as new certainties. Whatever conflicts may have arisen or been acerbated between elements of science and religion in, say, 1860, they were part of a much wider picture of change. And in the process of change, leaders of both religion and science have had to think seriously about what their new roles should be. Perhaps not enough.

Individual opinion always changes before authority. It is in the nature of authority to change slowly; society would be unstable otherwise. And religious authority may change slowest of all. The dilemma comes when change can no longer be put off. Furthermore, throughout history it has not simply been science that has driven societal change and nagged at ecclesiastical authority. There were always far broader forces at work in society, especially when more and more people were able to read and, for example, the Bible was available in languages other than Latin.

The period around 1860 marked a special period of challenge to authority, not only in terms of a conflict between science and religion but more broadly. This is demonstrated by looking, for example, at the problems experienced by the Church of England when confronted by change in secular society at exactly the period that Darwin's theory was launched, and largely independent of it.

In November 1861, some eighteen months after the Oxford Museum debate, Benjamin Disraeli, then Britain's chancellor of the exchequer (the equivalent in the United States of secretary of the treasury) in the second conservative government of Lord Derby, gave an address to a group of Anglican clerics in Oxfordshire. The speech was a classic case of authority resisting change because change is a messy thing to deal with and usually involves the simple becoming more complex.

In a typically stinging and attacking presentation, Disraeli took on the authors and the broad range of supporters of the book *Essays and Reviews*, with its revisionist views of a theology not based on miracles and of a science not based on the literal truth of the book of Genesis. Like Bishop Wilberforce, Disraeli was feeling embattled and needed to reinforce the authority of the Church of England—in this case not because of religious preference but from political necessity.

The unique role of the Church of England—the established church—in English civil society had been under continual attack from the 1820s onward, especially with the growth of the Midlands and northern industrial towns, where dissenting nonconformist groups flourished. Repeal of the Test and Corporation acts in 1828 had allowed nonconformists to hold public office for the first time. There was a similar Catholic Relief Act of 1829. And as of 1826, noncommunicants of the Church of England (banned from Oxford and Cambridge) had their own university—University College London.

With the election of growing numbers of dissenting members of Parliament, the church was losing its position of dominance in the House of Commons, even though its bishops occupied a powerful position in the House of Lords, where even today they occupy twenty-six seats—down from their maximum in Disraeli's time of ninety. (Who knew there were so many bishops?) Who knew—Disraeli did—what dangers would lie ahead if Anglican clergy and laymen started to act—and worse, vote—on the basis of having thought for themselves?

Judging from contemporary newspaper reports, Disraeli did not attack *Essays and Reviews* for its espousal of the new geology or Darwinian theory. He saw German biblical criticism as the principal source of disunity within the Church of England. If the churchmen were divided on the fundamental issues of scriptural interpretation and revelation, in all probability their sympathies would be equally strongly divided on social issues. So Disraeli tried to debunk the Reverend Rowland Williams's long essay on the subject of Christian Bunsen's researches on the biblical texts by subjecting the Germans as a whole to a piece of his famous withering rhetoric—always popular with an audience.

The speech was widely noticed in the press. As the *Aberdeen Journal* reported on November 20, 1861, Disraeli said,

> The volume in question [*Essays and Reviews*] is founded, generally speaking on the philosophical theology of Germany. . . . About a century ago, German philosophy, which was merely mysticism, became by a natural law of reaction, critical. A body of philosophical theologians accordingly arose and formed, in the course of years, a school which introduced a new system of interpretation of the Scriptures. They accepted without cavil the sacred narratives, but they explained all that was supernatural by natural causes. They adopted for the name of their new system the title of Rationalism. . . . Supported by great learning, and even greater ingenuity, the success of this school of philosophical theologians was transcendent. . . . But where is Rationalism, and where are the Rationalists? They have ceased to exist. . . . Another school of philosophical theologians arose in Germany, and with profound learning and inexorable logic, proved that Rationalism was irrational. They substituted for the rational scheme for interpreting the Scriptures a new scheme called the mythical . . . [just as] Rationalism was irrational . . . the mythical became a myth. . . . The new system is a most able revisal of pagan Pantheism.

By contrast, the liberal *Liverpool Mercury*, reporting the same speech on the same day, did not mention *Essays and Reviews* at all but focused on the immediate political issue: "Mr Disraeli . . . particularly labored to impress on the authority . . . the singularly mischievous fallacy that Churchmen can do anything they please with the legislation and government of the country if they can only agree among themselves. 'Depend upon it that nothing in this country can resist Churchmen when united.'"

The *Leeds Mercury*, another opposition newspaper, recognized Disraeli's theological remarks as pure electioneering:

"The Reverend Benjamin Disraeli," it sarcastically reported, "has been presiding at Aylesbury on the too much neglected Christian duty of setting people by the ears. . . . Looking back with longing eye to the time when a Tory ministry had an apparently perpetual tenure of office, he finds that there was then little or no fraternization between members of the Established Church and members of the Dissenting bodies. Mistaking a concomitant for a cause, Mr. Disraeli conceives that the best means to obtain a renewal of power for the Conservatives, is to renew that ancient sectarianism."

The *Aberdeen Journal* also reported on November 20, 1861, "On every subject on which he descanted, he was evidently more bent on talking eloquently than on giving any precise opinions. [Disraeli asked,] 'Is there anyone now who will say that the union between Church and State is not assailed and endangered. It is assailed in the principal place of the realm—its parliament; and it is in danger in an assembly where, if Churchmen were united, it would be irresistible' (cheers)."

The *Derby Mercury* set out some of Disraeli's specifics. "[In the last session of Parliament alone,] a series of bills were introduced . . . all converging on one point—an attack on the authority of the Church and on its most precious privileges. Your charities were assailed; even your churchyards were invaded. Your law of marriage was to be altered. Your public worship, to use the language of your opponents, was to be facilitated. Even the sacred fabrics of the churches were no longer to be considered national."

Disraeli was rightly worried about the church's monopoly over marriage practices having been diluted, a monopoly it had held since the Marriage Act of 1753. Before this act, couples could legitimately "marry" by a simple exchange of vows. After 1753, only weddings in a church by Church of England rite were valid, and that meant obtaining a license and the practice of "posting the banns" in advance. Everyone except Quakers and Jews was required to marry in the church until a compromise Marriage Act of 1823 made a "common-law" marriage without a license and banns legal; but then the minister who conducted the marriage was a felon. Finally, the Dissenters Marriage Act of 1836 allowed civil registrars to license a marriage conducted in either a licensed nonconformist chapel or a register office.

The next great change was that Parliament had established civil divorce courts. In England, divorce was almost unattainable before the passage of the Matrimonial Causes Act of 1857. Until then, divorce had been a complex, costly, and lengthy ecclesiastical proceeding in which neither party could be represented by civil lawyers, and each divorce required finding a sponsor for an individual Private Bill in Parliament. After 1857, there were hundreds of divorces each year; a floodgate had been eased open, and to traditionalists yet another bastion of the religious life—the sanctity of marriage and the marriage vows—had fallen.

Similarly, the Burial Act of 1859 allowed civil authority to take charge of church graveyards where there was a hazard to public health. The Church Rate Abolition Bill of 1859 allowed Dissenters and Catholics relief from a general tax that everyone previously had paid for the upkeep of Anglican churches.

With all these changes, science in the form of an un-proven Darwinism may have seemed a minor problem, but it gave yet another issue upon which the clergy could be (and were) divided. And looking back, it is not hard to see that Dis-raeli was right; this was a watershed period for the decline of religious authority in Britain.

One would have to reach all the way back to the late Middle Ages to find the sort of church-dominated society that would have suited Disraeli's political dreams. Then, in small towns and large, the church—both the edifice and the insti-tution that the vicar represented—was the center of civic life. The church was the source of moral and political guidance. The ecclesiastical courts held authority over issues from adul-tery to blasphemy and, as the church was also the major land-owner, over civil cases as well. In this the church was aided by the absence of general literacy, and the Bible in any case was in Latin, even for those who could afford one. Each vicar was a little king in his parish.

To exemplify the hold that the church had on society from medieval to post-Reformation times in England, we may visit the parish church of Wenhaston, Norfolk. Here, on the chancel arch that physically separates the realm of the priest from that of the congregation and symbolically marks the divide between the Church Expectant of redemption and everlasting life and the Church Militant of everyday life, there is a magnificent late medieval wall painting of the type known as a "doom." Dooms survive in churches all over England. The one at Wenhaston is not only a fine example dating to 1480, it also conjures up the wonderful novel *A Month in the Country* by J. L. Carr (1980), in which a similar painting in a remote Norfolk church is restored and all its gory detail revealed.

A doom is a depiction of Judgment Day and was placed where the people could see it. Indeed, they could not possibly avoid seeing it. It served to keep the largely illiterate faithful in line by reminding them of the coming day of reckoning. Christ, with the Virgin Mary and sometimes assisted also by St. Michael, stands at the upper center of a typical doom. He is weighing souls. Those found wanting and consigned to hell are seen in the painting exiting to his left, where they are shown suffering all manner of grotesque torture. Rather more weakly, the righteous move off to the opposite side toward an unending and undefined paradise. Subtlety is not a strong point of these images. Not for nothing are they called dooms (for sinners) rather than triumphs (for the faithful).

Even today, these paintings cause a shiver in the spine: painted scenes of hell with sinners being tortured, disemboweled, torn apart—fallen women and prosperous merchants all together in writhing masses of agony. When you see one of these for the first time, you recoil at the barbarity, not so much of the scene or even of the religious message, but of the minds that commissioned and authorized their placement in the most important part of the church, essentially to terrorize the congregation.

By the eighteenth century, the availability of cheap engravings made it possible for almost any home to have a copy of, for example, Michelangelo's *Day of Judgment* in the Sistine Chapel (1534–41) or Stefan Lochner's *Last Judgment* of 1435, where matters are laid out just as in the English dooms.

Benjamin Franklin, a deist like Jefferson, owned an image of the Day of Judgment (we don't know which one) "where the awful Judge was enthroned in glory and giving sentence; while some Souls were filing off to the right, and some, alas! To the left." For years it was stored in the attic. Then, when he lay dying, he had the picture brought down from storage. His nurse reported that "since he became poorly, and was confined to his bed, he requested her to bring it and place it at the foot of his bed, that he might have it always in his view . . . (and possibly) had some serious thought about the awful after Scene."[1]

A doom, with its message of promise, intimidation and, yes, knowledge, is a perfect example of authority at work, and its demise is a nice example of how everything always changes. The English Reformation removed all paintings from churches and no one sought later to revive them (except today as historical artifacts). In the 1840s Charles Darwin had rebelled at the idea of his father and friends being routinely consigned to hell. In a poll by the Pew Center for Religion and Public Life of 2008, while some 75–80 percent of Christians expressed a belief in heaven and assumed they are going there, only 60 percent believed in hell as a physical place.[2] The idea of hell is losing—perhaps has already lost—its terrors.

Religious authority, measured as the popular adherence to a particular set of beliefs, or as the vehicle for an institutional, ecclesiastical stance on issues like morality, has continued to decline over the last two millennia as versions of theology, of science, of society, and of religion have all changed. As knowledge of the Bible and, inevitably, interpretations of the concept of God's revelation have evolved (or at least become open to emendation), secular society has continued to evolve and so has science. Even God has evolved.

Some years ago, during a long flight across the United States, I absorbed myself in reading a book by Jack Miles. At one point I laid the book on my seat and took a walk to the back of the plane. When I returned, the flight attendant had it in her hands, its title clearly displayed—*God, a Biography.* And she asked me brightly, "This looks interesting, what is it about?" What could she have meant? The remarkable thing about this story is not that this young person might (quite impossibly) not have known who God was—rather it is her (and perhaps our) amazement that God has a biography, which implies change over time and perhaps (blasphemously) a life span, rather than just being. Yet "God's own book" documents how he has changed from the vengeful tyrant of the Old Testament, preoccupied with war and sex, to the God of love and forgiveness of the New. And just as our sense of God has evolved, so have other aspects of religious belief and practice.

One thing a historian would like to have is the results of opinion polls conducted in, say, the early nineteenth century. It is too bad there aren't any. But clearly there have been changes. Perhaps the biggest evolution in religious opinion that has occurred since, say, 1800, is that now only some 30–40 percent of the general population in the United States believe that the Bible is literally God's word, and 50 percent think it inspired but not literally true. In Jefferson's time the number thinking it literally true would have been close to 100 percent.

Public opinion polls conducted in recent years variously

by the Pew Research organization's Religion and Public Life Project and the Gallup, Bloomberg, and Barna organizations all suggest that somewhere between 85 percent and 92 percent of the U.S. population believe in God or a universal spirit.[3] A similar number (some 80 percent to 85 percent of the American population) accept that God was "the creator of the universe who rules the world today," and approximately the same number see God as personally concerned with each of us. These numbers change every year, and always in the same direction—away from the religious viewpoint. We no longer paint dooms in our churches, and hell does not hold us in thrall as it once did. Among American Christians, Satan has also suffered: 40 percent see him as nothing more than a symbol and, perhaps more surprisingly, only 35 percent see the Holy Spirit as a real entity rather than a metaphor.

Nothing, I suggest, should irritate modern scientists more than miracles. They were no less a thorn in the side of Jefferson, Hume, Paine, and every other one of our heroes of the Age of Reason. Yet it is quite remarkable that in this day and age, not only do a whopping 72 percent of the population believe that angels are a real phenomenon, some 80 percent believe in miracles.[4] The Catholic Church authenticates claims of miracles, and the performance of two miracles is a criterion for sainthood.

There can be no doubt of the intensity of conviction of those who claim to have witnessed miracles, unlikely as they may seem to others. There have not been many claims of miracles in the United States. On December 8, 1859, however, ironically within weeks of Darwin publishing *On the Origin of Species*, a Belgian immigrant woman to Wisconsin claimed to have been visited three times by the Virgin Mary, who "hovered between two trees in a bright light, clothed in dazzling white with a yellow sash around her waist and a crown of stars above her flowing blond locks" and instructed her to devote her life to teaching Catholic beliefs to children. One hundred and fifty years later, after a two-year investigation by theologians who found no evidence of fraud or heresy and a long history of shrine-related conversions, cures, and other signs of divine intervention, Bishop David L. Ricken of Green Bay declared "with moral certainty" that Ms. Brise did indeed have encounters "of a supernatural character" that are "worthy of belief."[5]

Interestingly, a 2009 Pew study of members of the American Association for the Advancement of Science found that 33 percent of scientists say they believe in God and another 18 percent believe in a universal spirit or higher power, adding up to a slight majority of 51 percent.[6] Overall, when you see these sorts of poll figures, you have to wonder why the world of religion is at all worried about science. Religion, in the United States, at least, seems to be doing just fine. And science not so well. Only some 40 percent of the general population accept that humans have arisen by evolution; 60 percent think that a worldwide flood accounts for most of the geological and paleontological record. Although only some 18 percent actually think the world is less than ten thousand years old, 50 percent believe the creation of life occurred in six literal days.[7]

As polls are repeated year by year, all the numbers continue to trend away from religion. In the past twenty years, the number of people considering religion to be important in public life has decreased from 88 percent to 80 percent. In Europe the numbers would be reversed—less than 20 percent thinking it important. The proportion of the U.S. population attending church regularly is down to 40 percent, and some analysts think it may really be only 20 percent. The Protestant churches have suffered the greatest decline.[8]

From an intellectual point of view, one of the most dispiriting things revealed by these exercises is that the Pew Survey of 2009 showed that only some 63 percent of the American population knew that Genesis is the first book of the Bible (and fewer than 50 percent of Roman Catholics did). Even more surprisingly, only 50 percent of those who identified themselves as Christians could name the four Gospels.[9] This should be just as depressing to the nonreligious public as it is to anyone else, because there is no place in society for ignorance. If religion and science are in conflict, then it would behoove both sides to know what they are talking about. And do children no longer at bedtime whisper, "Matthew, Mark, Luke, and John, guard the bed I lie upon?" Whatever you believe, the Bible is not only great literature (especially in the King James version); it is a core element of our culture. The same holds true for all the great foundational books of other religions.

If so many of the population of a nominally Christian country do not know what the Bible says, how can they judge whether or not to follow its precepts? I happen to think that the best hope for science in the United States would be for religion to be openly and frankly discussed in schools. Imagine religion debated under the Creationists' mantra of "Teach the controversy"! That is unlikely. We seem to be heading in the opposite direction. For example, in its political platform for 2012, the Republican Party of Texas stated: "We oppose the teaching of Higher Order Thinking Skills (values clarification), critical thinking skills and similar programs . . . which focus on behavior modification and have the purpose of challenging the student's fixed beliefs and undermining parental authority."[10] The party's mission is obviously to prevent, or at least minimize, both personal and public doubt. But doubt clearly is a fundamental ingredient of progress; it is the grease that keeps the wheels of society turning even when the cost is a temporary public dilemma.

CHAPTER TEN

A Way Forward?

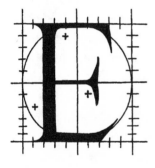VEN though religion (whatever we mean by that umbrella term) continues to be popular in America, and some 92 percent of the population professes to believe in God, it evidently continues to be worried about science. Some of this is obviously because religion in its various forms finds itself under attack both overtly and implicitly in secular society. There is, however, another side to the coin, in that science is coming more and more under attack by both Christians and Muslims. Those attacks are directed not only toward the teaching of evolution in schools but also, rather oddly, the subject of global climate change. This will not be the end of the list, partly because it reflects a general anti-intellectualism, a real fear of progressive thinking and a fear of the unknown, and resistance to challenges to conventional authority, all of which would have been perfectly

familiar to Jefferson's Federalist foes in the election of 1800. It exists, especially now, because science continues to reach more effectively into the realms of explanation of human behaviors such as morality and even the religious sense itself. (And that is certainly something we can blame Darwin for having started.) Interestingly, these opponents of science are far better supported financially and better organized than is science, the essence of which is to be dispersed and nonconformist.

One of the most dramatic and powerful changes of the last century was the insertion (some would say intrusion) of science into almost all aspects of public and private life. As government attempts to act for the most good for the most people, it creates a very different philosophy of society from Jefferson's vision of a nation of independent yeoman farmers. It enlists science and technology in support of everything from medicine, public health, the environment and the management of natural resources to public works, communications, and of course military applications. There seems to be no part of public life that is not now driven in large part by science.

In the decades after the Second World War and right through the 1970s, scholars and public leaders worried a great deal about the effect of science on government and the threat that it would become a power and an authority outside of the regular legislative processes. And scientists came to feel themselves under pressure as a result; the worth of their efforts was being questioned even though they then had little direct participation in the processes of government. But eventually, science came to be seen less as a bogeyman and, perhaps concomitantly, less as a solution to all problems. By and large scholars came to agree, if only through inertia, that science would not end up giving undue power to the executive or legislative branches of government but that the continuing problem would be whether the latter would use scientific authority responsibly.[1]

By and large, it seems that, by the end of the past century, we had arrived at a shifting balance with respect to this pervasive influence of science. As people came to understand that the world is infinitely complex and all parts are interrelated, we saw the need for more and more powerful tools for everything from space travel to medicine, even though the potential for misuse of the power that attaches to things like information management also grows and grows.

But everything continues to change, and one of the most significant developments of recent years is not the one-sided intrusion of science into government but, ironically enough, the rise of a religion-based politics striving for a similar position of authority. Things have long since passed beyond the stage of questioning the teaching of evolution and local school boards, as recent resistance to the role of a national standard of health care with respect to abortion shows. So it appears that where fifty years ago thoughtful scholars and shrill commentators questioned the influence of science on government, we have now to worry about an undue role of religion, not a religion that urges people to a more moral and caring life but one that is intellectually retrogressive—in Mr. Terry's terms, neither a broadened nor a purified religion.

All this thrusts us, however unwillingly, into yet another revival of the old antagonisms between religion and science as well as many new ones. That being the case, it becomes important to examine once again the nature and inevitability of those conflicts. While the basic elements of the debate have not moved away from the question of whether there is a God or not (and how, from a scientific point of view, we would know), everything else in the debate has become altered almost unrecognizably. Nonetheless, the question always arises: is there *necessarily* a conflict between science—or indeed any part of secular society—and religion?

The first answer—the quintessential academic's answer—has to be "That's the wrong question." Science and religion are but parts of a constantly changing nexus of local, national, and worldwide phenomena, some philosophical in nature, some intensely practical in effect. Science is very different when seen from (for example) a Muslim, Christian, or Hindu point of view, as is the relationship between religion and society.

Some modern historians of religion and science, led by John Hedley Brooke, tend to come close to denying that there is any *necessary* conflict at all.[2] They accuse their weaker brethren—historians of science, that is—of falling prey to what they call the naïve conflict myth: that science and religion have been engaged in a battle throughout history, one in which, unsurprisingly, science always "wins."

A favorite argument against the concept of *necessary* conflict is an appeal to the good old days when science and religion were not in conflict because all scientists were clerics. Those who deny the existence of a fundamental rift can fairly point out that there have been long periods when religious authority set the scientific agenda, and study of the natural world was more or less identifiable as the study of the properties, symmetries, orderliness, and the bounty and goodness of the mind of God. While this might technically be true, we might term this the "flat tire hypothesis." The tire is fine; it's only flat in one place. It also requires us to imagine that none of those early cleric-scientists—the intellectual foot soldiers (let alone the famous ones like Galileo)—ever came up with an observation or thought that was at odds with the religious authority under which they labored. Did no one doubt privately? As Keith Thomas has written elsewhere, "Not enough justice has been done to the volume of apathy, heterodoxy, and agnosticism which existed long before the onset of industrialism."[3]

In any case, science is not the same now as it was then—and nor is religion. And thank heaven for that. For a lot of the time before (and even well into) the scientific revolution, science was allied with magic and astronomers' (including Copernicus) carefully calculated horoscopes. Newton became preoccupied by alchemy. Now all of these things can be explained: Newton, for example, could be thought, in modern terms, to be experimenting with synthetic chemistry, horoscopes were a logical extension of Ptolemaic theory, and so on. And pious religious leaders thought it their duty to burn people alive to punish deviations from the contemporary orthodoxy.[4]

The second answer is that, yes, the conflict really is inherent and ongoing where it is based in differences that necessarily challenge authority. And it is in the very nature of things that science will always produce more "happenings" or developments that raise new questions for religious beliefs than the other way around. It is also in the very nature of things that, on average, authority changes more slowly than the innovative thinking that ultimately underpins it. We expect scientific authority to be quickly responsive to changes in knowledge. We all expect any religion, on the other hand, to the extent that it reflects and expresses the eternal verities (and certainly concepts that preceded Christianity), to change at a glacial pace. Where matters of practice and dogma affect a changing society, however, for example, in opinions on subjects like contraception or in vitro fertilization, we may fairly expect more adaptability.

Apart from denial, there are two other important approaches to the question of a conflict between religion and science. A postmodernist approach would be to reduce scientific facts and methods to the status of culturally mediated, if strongly held, opinions. They would then be less contentious but also less powerfully transformative. The search for a cure for cancer or HIV-AIDS would then be subject to the same level of nonscientific discourse as that to which religionists seek to reduce global climate change or, for some, to question the value of child vaccination. But while revisionists insist that both the sciences and conventional religions would both have to change radically, it turns that in most versions of the constructivist approach it is science that apparently needs to change most! In a fully constructivist model, to be fair, all religious beliefs would have to be reduced to opinion. It is not very likely that familiar religions, Christian or Muslim, would be willing to reduce their central notion of a supernatural God in such a way.

Perhaps the most popular hope for dealing with the differences between science and religion, over the millennia, has been the "independence model" favored by the Reverend Frederick Temple and so many others over history. As Samuel H. Thomson put it at a sermon at Hanover College in New Hampshire in 1858: "When you desire to know what God has revealed to us by his Spirit, look to the oracle of God. Look not to nature or to science. They are, indeed, full of truth; and that, too, which is God's truth, with which you cannot too carefully store your minds. But they are not the truths of revelation, and can never infallibly direct your eyes to them, nor take their place."[5] And in his "quasi-theology" sermon of 1878, Rev. Pusey said: "The basis of a lasting peace and alliance between physical science and Theology is, that neither should intrude into the province of the other. This is also true science. For science is *certain* knowledge based on *certain* facts. The facts on which Theology rests are spiritual facts: those of physical science are material."

In modern terms, this model of independence and mutual tolerance is usually expressed as religion and science being two different ways of knowing and therefore having separate realms of authority. Religion is the authority for a spiritual, moral world full of reverence, obligation, and purpose leading, for Christians, to final, eternal redemption by God's son. For science, it is the material world driven by Democritus's "chance and necessity" that ends with the sun petering out in some 5 billion years or so, by which time evolutionary events will have totally transformed life on earth anyway.

The late Stephen J. Gould—with, I suspect, only the very best of peacemaking ambitions—took this hoary old idea of "independence" and dressed it up with a special term, describing science and religion as *non-overlapping magisteria*.[6] His intent was finally to demonstrate that, since they were two distinct subjects, they ought to coexist peacefully, apart. Thus, as Professor Martin (chairman of the Terry Committee) has rightly said: science and religion can be seen as separate discourses. "It is not the job of Christian theology to question string theory. Science can say nothing about the basics of a belief in God or in Christ as a personal savior." Therefore, one might think that except for a few zealots, in terms of basic philosophy each side should leave the other alone, just as Darwin and so many others hoped.

A rigid application of Gould's idea of non-overlapping magisteria across the board, however, while a good example of "saving the phenomenon," is rather like Cinderella's sisters trying to force their feet all the way into the glass slipper. The conflict between religion and science exists not just with respect to debates about the nature and existence of God, or about the validity of any scientific idea, but *precisely because there are many places where the magisteria, the claims of authority of science and religion, do in fact overlap, intersect, and compete with each other.*

By the mid-nineteenth century, the facts of science had already invaded theological territory by degrading biblical authority concerning the origin and age of the earth. Science now seriously questions any supernatural involvement in the origin of life and demonstrates a lack of such involvement in the history of diversification of life on earth. It denies a universal Flood but provides evidence of a lesser flood local to the Levant that fits the biblical and other accounts (the Gilgamesh, for example) of a flood quite well (and in a way that the Reverend Pye Smith would have rejoiced to see). Science gives a whole different view of the origin and future evolution of the cosmos. Science does not afford humans an exceptional, supernaturally derived place in the evolution of life and currently is seriously engaged in attempts to show that features such as a moral sense and even a religious sense have an evolutionary origin (in some version of enlightened group self-interest consequent upon social structures).

These are all subjects in which various religions claim an authority largely or wholly consequent upon some sort of divine revelation or inspired teaching. And it is frankly very hard to see where any accommodation can be made safely to dilute the effects of science. That, no doubt, is why the idea of "separation" is so appealing—and dangerous.

A far more interesting side of the question of "magisteria" arises from the overlaps and from where conflicts are created by science in its more recent advances as it has extended its scope into areas so new that no clear-cut and explicit religious authority exists. These are areas where religious tenets lack a deep theological foundation, having only broad generalities in revelation for their basis. The majority of phenomena in this category involve serious questions in the moral-ethical sphere. I have in mind, for example, the whole range of bio-ethical issues inherent in biotechnology, stem cells, cloning, genetic engineering, ownership of cell lines, and practical issues like the economics (rationing) of health care, stewardship of the environment, nuclear power, global nuclear war, and global climate change. These issues do not concern theological abstractions or the supernatural but concern the conduct of daily life where science has an immediate as well as a principled impact. They say a great deal about how we act as humans but nothing about whether God created us or about evolution.

I cannot emphasize enough the existence and importance of this, still small, range of issues where both religion and science have "ownership" and religion has only shaky authority. Contraception is a good case in point. Perhaps no subject has caused as much dissension in the past one hundred years as contraception (and the broader subject of family planning), which has been the subject of repeated scientific and technical discoveries. Various religious groups condemn contraception, even making it a sin, while science not only continues to find new and safer means of contraception but also shows its benefits to society. Both sides have ownership of the issue. And here I will venture into contentious territory and state that the Bible offers no doctrinal guidance on which to make a case against contraception beyond the general "Go forth and multiply."[7]

If we consult any authoritative work, we find that the religious position against contraception depends to a very great extent on the "sin" of poor old Onan (Genesis 38), the progenitor, so to speak, of the withdrawal method. And there Onan was trying his best to act ethically because he had been ordered by Judah, his father, to impregnate his late brother's wife. For this fastidious discretion God killed him, as he had already killed Onan's brother Er, because "he was wicked in the sight of the Lord." (That was the old nasty God, but one can see why the withdrawal method might have its risks.) No other technique of family planning is unambiguously addressed by the Bible, and indeed there is no other method that is untainted by science. In the rhythm method, for example, how other than by the analysis of data does one discover what the rhythm is and how to use it?

Science has "ownership" of the contraception issue because it is the source of better and better, safer and safer, methods. And science puts its whole weight into matters of public health. When clerics and scientists discuss contraception, therefore, who addresses more effectively the tragedy that the leading cause of death in teenage girls, worldwide, is not disease but pregnancy?[8] The frequency of these deaths will not be reduced by pronouncements that, instead of distribution of condoms, we need "a humanization of sexuality, a human, spiritual renewal which brings with it a new way of behaving."[9] This quote comes from Pope Benedict XVI's proscription against the use of condoms in alleviating the spread of the HIV-AIDS virus in Africa.

And why not contraceptives? Because it is said to lead to a "banalization" of sex. Is there not something to be said for the equally banal utilitarian approach of doing the most good for the greatest number? I find it difficult to imagine justifying these children dying for lack of access to contraception on the basis of an abstract "banality" principle articulated by foreigners or, equally, national pride in population growth. If we are to go forth and multiply, which strategy has the best chance of success: allowing huge numbers of young mothers to die or guiding them safely to responsible and productive adulthood?

The debate over contraception is particularly interesting because, as science has made more and safer pharmacological approaches to the subject, adherence to existing religious pro-scriptions against contraception has rapidly declined. While there is a debate about the actual numbers, most studies show that more than half, and as many as 90 percent, of all Catholic women use some form of birth control other than the "natural method" at some time. Common sense, it seems, prevails. Once again, whenever new knowledge emerges, indi-viduals change faster than authority; private doubt creates a public dilemma.

The contemporary issues where science and religion have joint ownership are perhaps also the cases where, if doctrinal matters and prejudice (and sheer arrogance on all sides) can be put aside, religion and science can come together. Probably they are the only place where this can happen. Here, in my opinion, is the nub of the opportunity for present and future relations between religion and science: a mixture of a standoff in matters of belief, acceptance of facts, and cooperation in matters of public good. Is it hopelessly naïve to insist that we all, scientists, theologians, and the general public, should join forces to do good?

Perhaps our best chance to make a start is in environ-mental stewardship. On May 25, 2011, for example, an inter-esting group of people came forward in Philadelphia to testify at an Environmental Protection Agency hearing on mercury levels in the environment (there were similar hearings in other cities). And how unusual this combination was. There were, as one would expect, representatives of the Sierra Club, other environmental groups, physicians, industry representa-tives, unions, and private citizens but also the rabbinate, the Catholic Church, Protestant evangelicals, and Methodists.

These groups united not on grand sweeps of philosophy but on practical issues of power station emissions. The authority of the scientists was strengthened by the testimony of the religious groups; the cause of the religious groups was bolstered by the data and analyses of the scientists. In the middle were advocacy groups whose members included people of every persuasion but not necessarily any allegiance to either "side."

This is, if not a general model for forward action, at least an example of what might be possible. One can easily think of other subjects where science and religion, ethicists of every stripe, theologians, and captains of industry all have a common interest—issues definable as moral issues but only (or principally) realized because of advances in science and, arguably, much better resolved through cooperation and compromise than opposition. To put it simply, the opportunities lie in the overlaps between the so-called magisteria, not the separations.

I can summarize by saying that I believe that conflict between religion and science is not new, nor is it unique. It is part of a general phenomenon that necessarily happens when new knowledge meets the old. It starts with individuals—with each of us—and becomes more critical as systems of authority respond by changing at different rates. Of all systems, science and religion are almost guaranteed to change at different rates. But because science and religion are so important to so many people, and their spheres are coming more and more to overlap, we ought to be smart enough to deal with that.

I just wish I were optimistic that this will happen. I make that proviso because of the way that the conflict between religion and science is rising to the fore once more, this time because of the intrusion—or at least the extension—of a combination of anti-intellectualism and minority religious beliefs into the scientific sphere through politics. We have only to think of the recent contributions of two congressmen. Representative Todd Akin of Missouri believes that the female body has mechanisms so that rape cannot result in pregnancy. Congressman Paul Broun of Georgia (running unopposed in the past election) has said that evolution, embryology, and the big bang theory are "lies straight from the pit of hell . . . meant to convince people that they do not need a savior."[10] Why do we care about this? Well, both are members of the Science, Space and Technology Committee of the House of Representatives.

I mentioned earlier that for the authorities of the so-called Creation Research Institute behind the Creation Museum in Kentucky, "Accepting the Bible as God's literal truth doesn't mean that we discount science. It does mean that we interpret scientific evidence from the biblical viewpoint. . . . Evidence isn't labeled with dates and facts; we arrive at conclusions about the unobservable past based on our pre-existing beliefs." This approach (I hesitate to call it a philosophy) then empowers them to create slick museum exhibits claiming that the earth is no more than six thousand years old and depicting dinosaurs as living side by side with humans. They ignore the fossil record (some 65 million years separate the last dinosaur and the first *Homo sapiens*), using as justification a creative (so to speak) interpretation of the book of Job.

Job is a dubious source in the first place because Job is clearly only a folk narrative—or it had better be because it portrays God as such a despicable manipulator and vicious trickster. The supposed dinosaurs, the behemoth (Job 40:15) and leviathan (Job 41:1–34), have for more than a century been identified by serious independent biblical scholars as based on the hippopotamus and the crocodile respectively. They are probably best thought of as purely mythical, untamable beasts summoned out of the folkloric deeps to symbolize the power of God. Not dinosaurs. In any case, happily, the author of Job actually describes leviathan, which turns out not to be a very convincing real animal of any sort if "out of his mouth go burning lamps, and sparks of fire leap out. Out of his nostrils goeth smoke as out of a seething pot or caldron. His breath kindleth coals, and a flame goeth out of his mouth" (Job 41:19–21). That is a very comic-book kind of dinosaur worthy only of *Hagar the Horrible.*

Confronted with this sort of arrant and arrogant nonsense (an insult both to science and to the museum profession), most people just give a deep sigh and hope that it will all die a natural death, which may be far too relaxed a response. "Dinosaurs in Job" is not just a piece of ignorant foolishness; it is a highly calculated flagship challenge to the world of science and a raw contest for authority.

Unfortunately, we live in an age when this sort of thing thrives. Who, then, will speak for science? The rational viewpoint—scientists always claim the high ground, of course—is in danger of being left behind in a rush of politico-religious jockeying for authority. In the face of the shrill insistence that currently infests so much of the world, what is the solution? Obviously, the price of intellectual liberty is paid both in education and eternal vigilance. And that means that not just scientists but all people who could be termed "rational moderates"—of all stripes—must become more activist than at present. Scientists in particular, not just our leaders but the rank and file, need to engage more fully in the body politic.

The need for this marshaling of the forces of rationality is expressed perfectly in a statement I read recently and cheerfully borrow. *"We are called to be very active, very informed, and very involved in politics."*[11] This instruction comes from Cardinal Dolan, president of the United States Conference of Catholic Bishops, addressing those bishops in 2012. As a scientist I can only reply, AMEN. Yes, indeed we are—we all are called to be very active, very informed, and very involved in politics. And the overlaps between the spheres of influence of religion and science create a small window of opportunity for us to do so, carefully and deliberately, as colleagues and despite our differences, not as adversaries.

Bishop Samuel Wilberforce's Oxford Address

Bishop Samuel Wilberforce's address to the British Association for the Advancement of Science on June 30, 1860. (Freely imagined, largely on the basis of Wilberforce's published review of Darwin's *On the Origin of Species* in *Quarterly Review* [London], July 7, 1860, 225–64.)

My Lords, Ladies, and Gentlemen:

Our meeting here at Oxford's resplendent new museum, truly a temple of science and theology, has been a wonderful success. Presentations on dozens of subjects have been debated vigorously. Much of our discussion has been shadowed, I would not say, over-shadowed, by the recent work of Mr. Charles Darwin on the transmutation of species. The extent to which this idea has taken hold in some circles was demonstrated to you this morning in Professor Draper's presentation purporting to discover a species (if you will forgive the pun) of development theory applied to the laws and histories of nations. Professor Draper has been amply refuted by my colleague Mr. Cresswell and need not detain us longer. This morning I wish to address myself directly to the ideas of Mr. Darwin as contained in his volume *On the Origin of Species by Means of Natural Selection; or, The Preservation of Favoured Races in the Struggle for Life* — the title of which by itself gives the gist of his thesis.

Any contribution to our Natural History literature from the pen of Mr. C. Darwin is certain to command attention. His scientific attainments, his insight and carefulness as an observer, blended with no scanty measure of imaginative sagacity, and his clear and lively style, make all his writings unusually attractive. His present volume on the "Origin of Species" is the result of many years of observation, thought, and speculation; and is manifestly regarded by him as the "opus" upon which his future fame is to rest. It is true that he announces it modestly enough as the mere precursor of a mightier volume. But that volume is only intended to supply the facts which are to support the completed argument of the present essay. In this we have a specimen-collection of vast accumulation; and, working from these as the high analytical mathematician may work from the admitted results of his conic sections, he proceeds to deduce all the conclusions to which he wishes to conduct his readers.

The essay is full of Mr. Darwin's characteristic excellences. It is a most readable book; full of facts in natural history, old and new, of his collecting and of his observing; and all of these are told in his own perspicuous language, and all thrown into picturesque combinations, and all sparkle with the colours of fancy and the lights of imagination. It assumes, too, the grave proportions of a sustained argument upon a matter of the deepest interest, not to naturalists only, or even to men of science exclusively, but to every one who is interested in the history of man and of the relations of nature around him to the history and plan of Creation.

With Mr. Darwin's "argument" we may say in the outset that we shall have much and grave fault to find. But this does not make us the less disposed to admire the singular excellences of his work. Here, for instance, is a beautiful illustration of the wonderful interdependence of nature—of the golden chain of unsuspected relations which bind together all the mighty web which stretches from end to end of this full and most diversified earth. Who, as he listened to the musical hum of the great humble-bees, or marked their ponderous flight from flower to flower, and watched the unpacking of their trunks for their work of suction, would have supposed that the multiplication or diminution of their race, or the fruitfulness and sterility of the red clover, depend as directly on the vigilance of our cats as do those of our well-guarded game-preserves on the watching of our keepers? Yet this Mr. Darwin has discovered to be literally the case.

"From experiments which I have lately tried," he writes, "I have found that the visits of bees are necessary for the fertilisation of some kinds of clover; but humble-bees alone visit the red clover (Trifolium pratense), as other bees cannot reach the nectar. Hence I have very little doubt, that if the whole genus of humble-bees became extinct or very rare in England, the heartsease and red clover would become very rare or wholly disappear. The number of humble-bees in any district depends in a great degree on the number of field-mice, which destroy their combs and nests; and . . . the number of mice is largely dependent, as every one knows, on the number of cats; and . . . hence, it is quite credible that the presence of a feline animal in large numbers in a district might determine, through the intervention, first of mice, and then of bees, the frequency of certain flowers in that district."

Now, all this is, we think, really charming writing. We feel as we walk abroad with Mr. Darwin very much as the favoured object of the attention of the dervise must have felt when he had rubbed the ointment around his eye, and had it opened to see all the jewels, and diamonds, and emeralds, and topazes, and rubies, which were sparkling unregarded beneath the earth, hidden as yet from all eyes save those which the dervise had enlightened. But here we are bound to say our pleasure terminates; for, when we turn with Mr. Darwin to his "argument," we are almost immediately at variance with him. It is as an "argument" that the essay is put forward; as an argument we will test it.

We can perhaps best convey to this audience a clear view of Mr. Darwin's chain of reasoning, and of our objections to it, if we set out, first, the conclusion to which he seeks to bring them; next, the leading propositions which he must establish in order to make good his final inference; and then the mode by which he endeavours to support his propositions.

The conclusion, then, to which Mr. Darwin would bring us is, that all the various forms of vegetable and animal life with which the globe is now peopled, or of which we find the remains preserved in a fossil state in the great Earth-Museum around us, which the science of geology unlocks for our instruction, have come down by natural succession of descent

from father to son,—animals from at most four or five progenitors, and plants from an equal or less number, as Mr.
Darwin at first somewhat diffidently suggests; or rather, as,
growing bolder when he has once pronounced his theory, he
goes on to suggest to us, from one single head. "Analogy would
lead me one step further," Mr. Darwin proclaims, "namely,
to the belief that all animals and plants have descended from
some one prototype. But," he continues, "analogy may be a
deceitful guide." To which we may say, "Amen." "Nevertheless," he continues, "all living things have much in common
in their chemical composition, their germinal vesicles, their
cellular structure, and their laws of growth and reproduction.
. . . Therefore I should infer from analogy that probably all
the organic beings which have ever lived on this earth" (and
here Mr. Darwin has no hesitation in including man himself)
"have descended from some one primordial form into which
life was first breathed by the Creator."

This is the theory which really pervades the whole
volume. Man, beast, creeping thing, and plant of the earth,
are all the lineal and direct descendants of some one individual
ens, whose various progeny have been simply modified by the
action of natural and ascertainable conditions into the multi-
form aspect of life which we see around us. This is undoubt-
edly at first sight a somewhat startling conclusion to arrive at.
To find that mosses, grasses, turnips, oaks, worms, and flies,
mites and elephants, infusoria and whales, tadpoles of to-day
and venerable saurians, truffles and men, are all equally the
lineal descendants of the same aboriginal common ancestor,
perhaps of the nucleated cell of some primaeval fungus,
which alone possessed the distinguishing honour of being the

"one primordial form into which life was first breathed by the Creator"—this, to say the least of it, is no common discovery—no very expected conclusion. But we are too loyal pupils of inductive philosophy to start back from any conclusion by reason of its strangeness. Newton's patient philosophy taught him to find in the falling apple the law which governs the silent movements of the stars in their courses; and if Mr. Darwin can with the same correctness of reasoning demonstrate to us our fungular descent, we shall dismiss our pride, and avow, with the characteristic humility of philosophy, our unsuspected cousinship with the mushrooms.

Now, the main propositions by which Mr. Darwin's conclusion is attained are these: first that observed and admitted variations spring up in the course of descents from a common progenitor; secondly that many of these variations tend to an improvement upon the parent stock; and thirdly that, by a continued selection of these improved specimens as the progenitors of future stock, its powers may be unlimitedly increased. And, lastly, that there is in nature a power continually and universally working out this selection, and so fixing and augmenting these improvements. Mr. Darwin's whole theory rests upon the truth of these propositions, and crumbles utterly away if only one of them fail him. These therefore we must closely scrutinise.

We will begin with the last in our series, both because we think it the newest and the most ingenious part of Mr. Darwin's whole argument, and also because, whilst we absolutely deny the mode in which he seeks to apply the existence of the power to help him in his argument, yet we think that he throws great and very interesting light upon the fact that such a self-acting power does actively and continuously work in all Creation around us.

Mr. Darwin finds the disseminating and improving power, which he needs to account for the development of new forms in nature, in the principle of "Natural Selection," which is evolved in the strife for room to live and flourish which is evermore maintained between themselves by all living things. One of the most interesting parts of Mr. Darwin's volume is that in which he establishes this law of natural selection; we say establishes, because—repeating that we differ from him totally in the limits which he would assign to its action—we have no doubt of the existence or of the importance of the law itself. Mr. Darwin illustrates it thus:—"There is no exception to the rule that every organic being naturally increases at so high a rate, that, if not destroyed, the earth would soon be covered by the offspring of a single pair. Linnaeus has calculated that if an annual plant produced only two seeds—and there is no plant so unproductive as this—and their seedlings next year produced two, and so on, then in twenty years there would be a million plants. The elephant is reckoned the slowest breeder of all known animals, and I have taken some pains to estimate its probable minimum rate of natural increase. It will be under the mark to assume that it breeds when thirty years old, and goes on breeding till ninety years old, bringing forth three pair of young in this interval; if this be so, at the end of the fifth century there would be alive fifteen million elephants, descended from the first pair."

Now all this is excellent. Facts are gathered by Mr. Darwin from a true observation of nature, and from a patiently obtained comprehension of their undoubted and unquestionable relative significance. That such a struggle for life then actually exists, and that it tends continually to lead the strong to exterminate the weak, we readily admit; and in this law we see a merciful provision against the deterioration, in a world apt to deteriorate, of the works of the Creator's hands. Thus it is that the bloody strifes of the males of all wild animals tend to maintain the vigour and full development of their race; because, through this machinery of appetite and passion, the most vigorous individuals become the progenitors of the next generation of the tribe. And this law, which thus maintains through the struggle of individuals the high type of the family, tends continually, through a similar struggle of species, to lead the stronger species to supplant the weaker.

But this indeed is no new observation. We remember from our schooldays that Lucretius knew and eloquently expatiated on its truth: — "Multaque tum interiisse animantum secla necesse est / Nec potuisse propagando procudere prolem / Nam, quaequomque vides vesci vitalibus auris / Aut dolus, aut virtus, aut denique mobilitas, est / Ex ineunte aevo, genus id tutata reservans."[1]

And this, which is true in animal, is no less true in vegetable life. Hardier or more prolific plants, or plants better suited to the soil or conditions of climate, continually tend to supplant others less hardy, less prolific, or less suited to the conditions of vegetable life in those special districts. Thus far, then, the action of such a law as this is clear and indisputable.

But before we can go a step further, and argue from its operation in favour of a perpetual improvement in natural types, we must be shown first that this law of competition has in nature to deal with such favourable variations in the individuals of any species, as truly to exalt those individuals above the highest type of perfection to which their least imperfect predecessors attained—above, that is to say, the normal level of the species, and then, next, we must be shown that there is actively at work in nature, co-ordinate with the law of competition and with the existence of such favourable variations, a power of accumulating such favourable variation through successive descents. Failing the establishment of either of these last two propositions, Mr. Darwin's whole theory falls to pieces.

Mr. Darwin begins by endeavouring to prove that such variations are produced under the selecting power of man amongst domestic animals. Now here we demur *in limine*. Mr. Darwin himself allows that there is a plastic habit amongst domesticated animals which is not found amongst them when in a state of nature. He writes a delightful chapter upon pigeons. Runts and fantails, short-faced tumblers and long-faced tumblers, long-beaked capriers and pouters, black barbs, jacobins, and turbits, coo and tumble, inflate their oesophagi, and pout and spread out their tail before us. We learn that "pigeons have been watched and tended with the utmost care, and loved by many people." They have been domesticated for thousands of years in several quarters of the world. The earliest known record of pigeons is in the fifth Egyptian dynasty, about 3000 B.C., though "pigeons are given in a bill of fare" (what an autograph would be that of the chef-de-cuisine of the day!) "in the previous dynasty." Mr. Darwin quotes with approval Sir John Sebright, who has apparently said, with respect to breeding pigeons, that "he would produce any given feather in three years, but it would take him six years to produce beak and head."

Now all this is again very pleasant writing, especially for pigeon-fanciers; but what step do we really gain in it at all towards establishing the alleged fact that variations are but species in the act of formation, or in establishing Mr. Darwin's position that a well-marked variety may be called an incipient species? We affirm positively that no single *fact* tending even in that direction is brought forward. On the contrary, every one points distinctly towards the opposite conclusion; for with all the change wrought in appearance, with all the apparent variation in manners, there is not the faintest beginning of any such change.

There is no tendency to that great law of sterility which, in spite of Mr. Darwin, we affirm ever to mark the hybrid; for every variety of pigeon, and the descendants of every such mixture, breed as freely, and with as great fertility, as the original pair; nor is there the very first appearance of that power of accumulating variations until they grow into specific differences, which is essential to the argument for the transmutation of species; for, as Mr. Darwin allows, sudden returns in colour, and other most altered appearances, to the parent stock continually attest the tendency of variations not to become fixed, but to vanish, and manifest the perpetual presence of a principle which leads not to the accumulation of minute variations into well-marked species, but to a return from the abnormal to the original type. So clear is this, that it is well known that any relaxation in the breeder's care effaces all the established points of difference, and the fancy-pigeon reverts again to the character of its simplest ancestor.

There is another race of animals which comes under our closest inspection, which has been the friend and companion of man certainly ever since the wandering Ulysses returned to Ithaca, and of which it has been man's interest to obtain every variation which he could extract out of the original stock. The result is every day before us. We all know the vast difference, which strikes the dullest eye, between, for instance, the short bandylegged snub-nosed bull-dog, and the almost aerial Italian gray-hound. Here again the experiment of variation by selection has been well-nigh tired out. And with what results? Here again with an absolute absence of the first dawns of any variety which could by its own unlimited prolongation constitute a specific difference. Again there is perfect freedom and fertility of interbreeding; again a continual tendency to revert to the common type; again, even in the most apparently dissimilar specimens, a really specific agreement. Nor should we forget over how large a lapse of time our opportunities of observation extend. From the early Egyptian habit of embalming, we know that for 4000 years at least the species of our own domestic animals, the cat, the dog, and others, has remained absolutely unaltered.

Nor must we pass over unnoticed the transference of the argument from the domesticated to the untamed animals. Assuming that man as the selector can do much in a limited time, Mr. Darwin argues that Nature, a more powerful, a more continuous power, working over vastly extended ranges of time, can do more. But why should Nature, so uniform and persistent in all her operations, tend in this instance to change? why should she become a selector of varieties? Because, most ingeniously argues Mr. Darwin, in the struggle for life, *if* any variety favourable to the individual were developed, that individual would have a better chance in the battle of life, would assert more proudly his own place, and, handing on his peculiarity to his descendants, would become the progenitor of an improved race; and so a variety would have grown into a species.

We think it difficult to find a theory fuller of assumptions; and of assumptions not grounded upon alleged facts in nature, but which are absolutely opposed to all the facts we have been able to observe.

The applied argument then, from variation under domestication, fails utterly. But further, what does observation say as to the occurrence of a single instance of such favourable variation? Men have now for thousands of years been conversant as hunters and other rough naturalists with animals of every class. Has any one such instance ever been discovered? We fearlessly assert not one. Variations have been found: rodents whose teeth have grown abnormally; animals of various classes of which the eyes, from the absence of light in their dwellings, have been obscured and obliterated; but *not one* which has tended to raise the individual in the struggle of life above the typical conditions of its own species. Mr. Darwin himself allows that he finds none; and accounts for their absence in existing fauna only by the suggestion, that, in the competition between the less improved parent-form and the improved successor, the parent will have yielded in the strife in order to make room for the successor; and so "both the parent and all the transitional varieties will generally have been exterminated by the very process of formation and perfection of the new form,"—a most unsatisfactory answer as it seems to us; for why—since if this is Nature's law these innumerable changes must be daily occurring—should there never be any one produceable proof of their existence?

Here then again, when subjected to the stern Baconian law of the observation of facts, the theory breaks down utterly; for no natural variations from the specific type favourable to the individual from which nature is to select can anywhere be found.

But once more: if these transmutations were actually occurring, must there not, in some part of the great economy of nature round us, be somewhere at least some instance to be quoted of the accomplishment of the change? With many of the lower forms of animals, life is so short and generations so rapid in their succession that it would be all but impossible, if such changes were happening, that there should be no proof of their occurrence; yet never have the longing observations of Mr. Darwin and the transmutationists found one such instance to establish their theory, and this although the shades between one class and another are often most lightly marked. For there are creatures which occupy a doubtful post between the animal and the vegetable kingdoms—half-notes in the great scale of nature's harmony. Is it credible that all favourable varieties of turnips are tending to become men, and yet that the closest microscopic observation has never detected the faintest tendency in the highest of the Algae to improve into the very lowest Zoophyte?

Again, we have not only the existing tribes of animals

out of which to cull, if it were possible, the instances which the transmutationists require to make their theory defensible consistently with the simplest laws of inductive science, but we have in the earth beneath us a vast museum of the forms which have preceded us. Over so vast a period of time does Mr. Darwin extend this collection that he finds reasons for believing that "it is not improbable that a longer period than 300,000,000 years has elapsed since the latter part of the secondary [geological] period" alone. Here then surely at last we must find the missing links of that vast chain of innumerable and separately imperceptible variations, which has convinced the inquirer into Nature's undoubted facts of the truth of the transmutation theory. But no such thing. The links are wholly wanting, and the multiplicity of these facts and their absolute rebellion against Mr. Darwin's theory is perhaps his chief difficulty.

The other solvent which Mr. Darwin most freely and, we think, unphilosophically employs to get rid of difficulties, is his use of time. This he shortens or prolongs at will by the mere wave of his magician's rod. Thus the duration of whole epochs, during which certain forms of animal life prevailed, is gathered up into a point, whilst an unlimited expanse of years, impressing his mind with a sense of eternity, is suddenly interposed between that and the next series, though geology proclaims the transition to have been one of gentle and, it may be, swift accomplishment. All this too is made the more startling because it is used to meet the objections drawn from facts. "We see none of your works," says the observer of nature; "we see no beginnings of the portentous change; we see plainly beings of another order in Creation, but we find amongst them no tendencies to these altered organisms." True, says the great magician, with a calmness no difficulty derived from the obstinacy of facts can disturb; "true, but remember the effect of time. Throw in a few hundreds of millions of years more or less, and why should not all these changes be possible, and, if possible, why may I not assume them to be real?" Together with this large licence of assumption we notice in this book several instances of receiving as facts whatever seems to bear out the theory upon the slightest evidence, and rejecting summarily others, merely because they are fatal to it. We grieve to charge upon Mr. Darwin this freedom in handling facts, but truth extorts it from us. That the loose statements and unfounded speculations of this book should come from the author of the monographs on Cirripedes, and the writer, in the natural history of the Voyage of the "Beagle," of the paper on the Coral Reefs, is indeed a sad warning how far the love of a theory may seduce even a first-rate naturalist from the very articles of his creed.

But not only do the facts to which Mr. Darwin trusts to establish his vast lapses of years, which, he says, "impress his mind almost in the same manner as does the vain endeavour to grapple with the idea of Eternity," not only do these give him the same power of supposing the progress of changes, of which we have found neither the commencement, nor the progress, nor the record, as ancient geographers allowed themselves, when they speculated upon the forms of men whose heads grew beneath their shoulders in the unreached recesses of Africa,—but when, passing from these unlimited terms for change to work in, he proceeds to deal with the absence of all record of the changes themselves, the plainest geological facts again disprove his assumptions. Yet Mr. Darwin is compelled to admit that he finds no records in the crust of the earth to verify his assumption. "To the question why we do not find records of these vast primordial periods," he writes, "I can give no satisfactory answer."

We come then to these conclusions. All the facts presented to us in the natural world tend to show that none of the variations produced in the fixed forms of animal life, when seen in its most plastic condition under domestication, give any promise of a true transmutation of species; first, from the difficulty of accumulating and fixing variations within the same species; secondly, from the fact that these variations, though most serviceable for man, have no tendency to improve the individual beyond the standard of his own specific type, and so to afford matter, even if they were infinitely produced, for the supposed power of natural selection on which to work; whilst all variations from the mixture of species are barred by the inexorable law of hybrid sterility. Further, the embalmed records of 3000 years show that there has been no beginning of transmutation in the species of our most familiar domesticated animals; and beyond this, that in the countless tribes of animal life around us, down to its lowest and most variable species, no one has ever discovered a single instance of such transmutation being now in prospect; no new organ has ever been known to be developed—no new natural instinct to be formed—whilst, finally, in the vast museum of departed animal life which the strata of the earth imbed for our examination, whilst they contain far too complete a representation of the past to be set aside as a mere imperfect record, yet afford no one instance of any such change as having ever been in progress, or give us anywhere the missing links of the assumed chain, or the remains which would enable now existing variations, by gradual approximations, to shade off into unity.

On what then is the new theory based? We say it with unfeigned regret, in dealing with such a man as Mr. Darwin, on the merest hypothesis, supported by the most unbounded assumptions.

You will not have failed to notice that we have objected to the views with which we have been dealing solely on scientific grounds. We have done so from our fixed conviction that it is thus that the truth or falsehood of such arguments should be tried. We have no sympathy with those who object to any facts or alleged facts in nature, or to any inference logically deduced from them, because they believe them to contradict what it appears to them is taught by Revelation. We think that all such objections savour of a timidity which is really inconsistent with a firm and well-instructed faith.

In his *Discourse on the Studies of the University*, Professor Sedgwick, a man of deep thought and great practical wisdom, one whose piety and benevolence have for many years been shining before the world, and of whose sincerity no scoffer (of whatever school) will dare to start a doubt, recorded his opinion "that Christianity had everything to hope and nothing to fear from the advancement of philosophy."

This is as truly the spirit of Christianity as it is that of philosophy. Few things have more deeply injured the cause of religion than the busy fussy energy with which men, narrow and feeble alike in faith and in science, have bustled forth to reconcile all new discoveries in physics with the word of inspiration. For it continually happens that some larger collection of facts, or some wider view of the phenomena of nature, alter the whole philosophic scheme; whilst Revelation has been committed to declare an absolute agreement with what turns out after all to have been a misconception or an error. We cannot, therefore, consent to test the truth of natural science by the Word of Revelation. But this does not make it the less important to point out on scientific grounds scientific errors, when those errors tend to limit God's glory in Creation, or to gainsay the revealed relations of that Creation to Himself. To both these classes of error, though, we doubt not, quite unintentionally on his part, we think that Mr. Darwin's speculations directly tend.

Mr. Darwin writes as a Christian, and we doubt not that he is one. We do not for a moment believe him to be one of those who retain in some corner of their hearts a secret unbelief which they dare not vent; and we therefore pray him to consider well the grounds on which we brand his speculations with the charge of such a tendency. First, then, he not obscurely declares that he applies his scheme of the action of the principle of natural selection to MAN himself, as well as to the animals around him.

Now, we must say at once, and openly, that such a notion is absolutely incompatible not only with single expressions in the word of God on that subject of natural science with which it is not immediately concerned, but, which in our judgment is of far more importance, with the whole representation of that moral and spiritual condition of man which is its proper subject-matter. Man's derived supremacy over the earth; man's power of articulate speech; man's gift of reason; man's free-will and responsibility; man's fall and man's redemption; the incarnation of the Eternal Son; the indwelling of the Eternal Spirit,—all are equally and utterly irreconcilable with the degrading notion of the brute origin of him who was created in the image of God, and redeemed by the Eternal Son assuming to himself his nature.

Equally inconsistent, too, not with any passing expressions, but with the whole scheme of God's dealings with man as recorded in His word, is Mr. Darwin's daring notion of man's further development into some unknown extent of powers, and shape, and size, through natural selection acting through that long vista of ages which he casts mistily over the earth upon the most favoured individuals of his species.

Nor can we doubt, secondly, that this view, which thus contradicts the revealed relation of Creation to its Creator, is equally inconsistent with the fulness of His glory. It is, in truth, an ingenious theory for diffusing throughout Creation the working and so the personality of the Creator. And thus, however unconsciously to him who holds them, such views really tend inevitably to banish from the mind most of the peculiar attributes of the Almighty.

How, asks Mr. Darwin, can we possibly account for the manifest plan, order, and arrangement which pervade Creation, except we allow to it this self-developing power through modified descent? How indeed can we account for all this? By the simplest and yet the most comprehensive answer. By declaring the stupendous fact that all Creation is the transcript in matter of ideas eternally existing in the mind of the

Most High—that order in the utmost perfectness of its relation pervades His works, because it exists as in its centre and highest fountain-head in Him the Lord of all. Shall we, instead, learn from the apes?

I spy Professor Huxley in the audience, I wonder if he . . .

Notes

PREFACE

1. *Charles Darwin's Notebooks, 1836–1844*, ed. Paul H. Barrett, Peter J. Gautry, Sand Herbert, David Kohn, and Sydney Smith (Ithaca: Cornell University Press, 1987), M Notebook, entry 40. The full entry reads: "There is absolute pleasure independent of imagination, (as in *hearing* music), this probably arises from (1) harmony of colours & their absolute beauty (which is as real a cause as in music) from the splendour of light. . . . Again there is beauty in rhythm & symmetry . . . connection with poetry, abundance, fertility, rustic life, virtuous happiness— recall scraps of poetry . . . recall pictures & therefore imagining pleasure." The entry concludes: "I a geologist have ill defined notion of land covered with ocean, former animals, slow force cracking surface &c truly poetical (v. Wordsworth about science being sufficiently habitual to become poetical."

2. *Charles Darwin's Notebooks*, N Notebook, entry 57.

3. William Wordsworth, *Selected Poems and Prefaces*, ed. Jack Stillinger (Boston: Houghton Mifflin, 1965), 455–456.

CHAPTER I. THE LONG-STANDING PROBLEM

1. *Essays and Reviews* (London: John W. Parker, 1860).

2. Sometimes when I read the words "science" and "religion" linked together, I am reminded of the German playwright Hanns Johst: "Wenn ich Kultur höre . . . entsichere ich meinen Browning!" (Whenever I hear of culture . . . I release the safety catch of my Browning!) *Schlageter* act 1, scene 1 (1933).

3. *The Autobiography of Benjamin Rush: His "Travels through Life" Together with*

His Commonplace Book for 1789–1813, ed. George Corner (Philadelphia: American Philosophical Society, 1948).

4. "Science," *Annual Register* 113 (1871): 368.

5. Christopher Lane, *The Age of Doubt: Tracing the Roots of Our Religious Uncertainty* (New Haven: Yale University Press, 2011).

6. *The Religion and Science Debate: Why Does It Continue?* ed. Harold W. Attridge (New Haven: Yale University Press, 2009).

7. J. Arthur Thomson, *Concerning Evolution* (New Haven: Yale University Press, 1925).

8. Edith Hamilton, *The Greek Way* (New York: Norton 1942), 16.

9. The Reverend Gilbert White in his *History and Antiquities of Selborne* (1789) followed medieval naturalists in believing that swallows overwinter in ponds, even when his brother gave him evidence that they migrate. More scientifically, he set out to test whether bees, as Virgil had said in *The Georgics*, could hear. He shouted at his hives with a speaking trumpet and they ignored him. My own mother, however, whenever there was a major family event like a birth or death, would go to our row of hives to "tell the bees." The underlying reason in folklore (although she never articulated it that way) was that bees spread out over the countryside and would carry the news with them. She probably didn't know why she told the bees; it was simply something that everyone in her rural upbringing had done. It is not a practice that I have continued. The romantic in me finds a tiny little loss in that, while the rationalist would not have it any other way.

10. We have a serious terminological problem in dealing with evolution and many other topics. We commonly say that we "believe in" or do not "believe in" Darwinian evolution, which reduces the matter apparently to opinion. What we mean is that "we accept the facts of" or "do not accept the facts of" evolution. We do not say, for example, that we "believe" Newton's three laws of motion.

11. John Banville, *Dr. Copernicus* (London: Secker and Warburg, 1976), 29.

CHAPTER 2. RELIGION AND SCIENCE

1. Studies of science in all its internal and external contexts abound. Of all that has been written on the nature of science and scientific method, among the most accessible commentaries are now quite venerable works: Peter B. Medawar, *Induction and Intuition in Scientific Thought* (Philadelphia: American Philosophical Society, 1969); Thomas S. Kuhn, *The Structure of Scientific Revolutions* (Chicago: University of Chicago Press, 1962); and W. H. Newton-Smith, *The Rationality of Science* (Boston: Routledge and Kegan Paul, 1981). Important recent source books for discussion of the relationships between religion and science include Thomas Dixon, Geoffrey Cantor, and Stephen Pumprey, eds., *Science and Religion: New Historical Perspectives* (Cambridge: Cambridge University Press, 2010);

and Peter Harrrison, ed., The *Cambridge Companion to Science and Religion* (Cambridge: Cambridge University Press, 2010).

2. Keith Thomson, "The Revival of Experiments on Prayer," *American Scientist* 84 (1996): 523–34.

3. Marcus Tullius Cicero, *De Natura Deorum; or, On the Nature of the Gods*, ed. H. Rackham (New York: Loeb Classical Library, 1933), 87–88. David Hume used the same rhetorical strategy to examine the same questions in his *Dialogues concerning Natural Religion*, published posthumously in 1779.

4. Cicero, *De Natura Deorum*, 93–94. Of course, in *De Divinatione*, Cicero also said, "There is nothing so ridiculous but some philosopher has said it."

5. John Polkinghorne, *Belief in God in an Age of Science* (New Haven: Yale University Press, 1998), 6.

6. Ibid., 9.

7. Not to be facetious, but dogs actually think the same. That is, they have a totally different way from us of physically perceiving the world and the intelligence needed to process it. For them, the universe is largely a three-dimensional array of odors, and humans are simply a useful by-product of the current stage of their evolution—one that makes it no longer necessary for them to forage for food, for example.

8. In fact, I have to admit that my wife—in a letter to the *Times* of London, no less, which never accepts *my* letters—once described me as pedantic. So be it.

9. Clifford Geertz, *The Interpretation of Cultures* (New York: Basic Books, 1973), 90.

10. Keith Thomson, introduction to Attridge, *The Religion and Science Debate*, 4–5.

11. David Hume, *An Enquiry concerning Human Understanding* (Chicago: Gateway Edition, 1956), p. 173.

12. *New York Times*, October 23, 2011.

CHAPTER 3. MR. JEFFERSON'S DILEMMA

1. Frank Shuffletton's edition of Thomas Jefferson's *Notes on the State of Virginia* contains an excellent introduction to Jefferson's scientific thinking (New York, Penguin, 1999), vii–xxxvi.

2. F. J. de B. Chevalier de Chastellux, *Travels in North America in the Years 1780, 1781, and 1782*, ed. Howard C. Rice (Chapel Hill: University of North Carolina Press, 1963), 2:393–95.

3. Authorship uncertain; see Richard M. Dorson, "The Jonny-Cake Papers" *Journal of American Folklore* 58 (1945): 104.

4. Keith Thomson, *Jefferson's Shadow: The Story of His Science* (New Haven: Yale University Press, 2012).

5. Thomas Jefferson to John Adams, April 11, 1823, in *The Adams-Jefferson Letters: The Complete Correspondence between Thomas Jefferson & Abigail & John Adams*, ed. Lester J. Capon (Chapel Hill: University of North Carolina Press, 1959), 592. Note that by "without appeal to revelation," he meant on no authority other than his own reason.

6. Isaac Newton, *The Principia: Mathematics of Natural Philosophy*, ed. I. Bernard Cohen and Anne Whitman (Berkeley: University of California Press, 1999), 940.

7. Georges-Louis Leclerc, Comte de Buffon, *Histoire naturelle, générale et particulière, avec déscription du cabinet du roi*, vol. 9 (Paris: Imprimerie Royale, 1761), 101-10.

8. Jefferson to Charles Thomson, December 17, 1786, in *The Papers of Thomas Jefferson*, ed. J. Boyd (Princeton: Princeton University Press, 1954), 10:608-10. Thomson made the first translation of the Bible in America and was an expert on American Indians, contributing an important section on that subject to Jefferson's *Notes on the State of Virginia*.

9. Thomas Burnet, *Telluria Theoria Sacra: Sacred History of the Earth* (London, 1681), ed. Basil Wiley (London: Centaur, 1965).

10. Jefferson to Charles Thomson, September 20, 1787, in *Papers*, 12:159-61.

11. Thomas Jefferson, *Notes of a Tour into the Southern Parts of France*, in *Papers*, 11:460-62.

12. Keith Thomson, *A Passion for Nature; Thomas Jefferson and Natural History* (Chapel Hill: University of North Carolina Press, 2008).

13. Keith Thomson, "The 'Great-Claw' and the Science of Thomas Jefferson," *Proceedings of the American Philosophical Society* 155 (2010): 394-403.

14. An amusing opposite view is found in a letter that Benjamin Smith Barton wrote to Jefferson in 1811 noting that the English Methodist theologian Adam Clark, in a commentary on the Bible, suggested that "in beasts God shows his wondrous Skill and power: in the vast elephant, and Still more the colossal mammoth, or megalonyx the whole race of which seems to be extinct. . . . He seems to have produced him merely to show what he could do." Barton to Jefferson, February 1, 1811, in *Papers, Retirement Series*, 3:356-57. There is no reason to suppose that Jefferson found this either compelling or amusing.

15. Charles Thomson to Jefferson, April 28, 1787, in *Papers*, 11:323-24.

16. Jefferson to Charles Thomson, September 20, 1787, in *Papers*, 12:159-61.

17. Jefferson, *Notes on the State of Virginia*, 21.

18. Clement C. Moore, *Observations upon Certain Passages in Mr. Jefferson's Notes on Virginia* (New York, 1804), 30-31.

19. Timothy Dwight, *The Duty of Americans, at the Present Crisis, Illustrated in a Discourse Preached on the Fourth of July, 1798* (New Haven, 1798), 20.

20. Joseph J. Ellis, *American Sphinx: The Character of Thomas Jefferson* (New York: Vintage Books, 1996).

21. Charles Miller, *Jefferson and Nature* (Baltimore: Johns Hopkins University Press, 1988), 46, 48.

22. Jefferson to C. F. C. de Volney, February 8, 1805, in *The Writings of Thomas Jefferson*, ed. Andrew A. Lipscomb and Albert Ellery Berg (Washington, D.C.: Thomas Jefferson Memorial Foundation, 1903-4), 11:62-69.

23. Jefferson to Dr. John P. Emmett, May 2, 1826, in *Writings*, 16:168-72.

CHAPTER 4. ANCIENT OF DAYS

1. The most comprehensive account of early modern geological thinking is contained in the following pair of volumes: Martin J. S. Rudwick, *Bursting the Limits of Time* (Chicago: University of Chicago Press, 2005) and *Worlds before Adam* (Chicago: University of Chicago Press, 2008). See also Rachel Laudan, *From Mineralogy to Geology: The Foundations of a Science, 1650-1830* (Chicago: University of Chicago Press, 1987); and Mott T. Greene, *Geology in the Nineteenth Century: Changing Views of a Changing World* (Ithaca: Cornell University Press, 1982).

2. Baden Powell, *The Order of Nature Considered in Reference to the Claims of Revelation* (London: Longman, 1859), 443.

3. Ibid., 6.

4. James Hutton, "The Theory of the Earth," *Transactions of the Royal Society of Edinburgh* 1 (1788): 209-304.

5. Chalmers became a professor of theology at St. Andrews. In 1814 he published a revised version of his thesis: "Remarks on Cuvier's Theory of the Earth," in *The Works of Thomas Chalmers, D.D. & L.L.D* (Glasgow: Collins, 1840), 12:349-72.

6. Another solution proposed by philologists was that the word *and* at the beginning of the second sentence was wrongly translated, producing a linkage that was not intended. In that case, the condition of the earth "without form, and void" could even have been the result of a catastrophe that destroyed some earlier version(s) of the earth.

7. Benjamin Silliman, in Robert Bakewell, *An Introduction to Geology*, 3rd ed. (New Haven: Howe, 1829), 67.

8. Ibid., 67-68.

9. Benjamin Silliman, *Suggestions Relative to the Philosophy of Geology as Deduced from the Facts & to the Consistency of Both the Facts & Theory of This Science with Sacred History* (New York: Hamlen, 1839), 101.

10. John H. Glitner, *Moses Stuart: The Father of Biblical Science in America* (Atlanta: Scholars, 1988).

11. Quoted in Conrad Wright, "The Religion of Geology," *New England Quarterly* 14 (1941): 335–58.

12. John Pye Smith, *The Relation between the Holy Scriptures and Some Parts of Geological Science*, 4th ed. (London: Peterson, 1843), 147.

13. Ibid., 198.

14. Ibid., 197–200.

15. Charles H. Hitchcock, *The Relations of Geology to Theology* (Andover: Draper, 1867).

16. Hugh Miller, *Footprints of the Creator; or, The Asterolepis of Stromness*, 3rd ed. (Boston: Gould and Lincoln, 1863), 330–31.

17. Jefferson to John Adams, January 24, 1814, in *Adams-Jefferson Letters*, 421.

18. William Ewart Gladstone, "Days of Creation and of Worship," in [no author] *The Order of Creation: The Conflict between Genesis and Geology* (New York: Truth Seeker, n.d.), 25 (a compilation of letters by Gladstone, Huxley, and others).

CHAPTER 5. MR. DARWIN'S RELIGION

1. Of the many comprehensive biographies of Charles Darwin, two stand out: Janet Browne, *Charles Darwin*, vol. 1, *Voyaging* (New York: Knopf, 1995) and vol. 2, *The Power of Place* (New York: Knopf, 2002); and Adrian Desmond and James Moore, *Darwin: The Life of a Tormented Evolutionist* (New York: Norton, 1991).

2. Benjamin Disraeli, *Tancred; or, The New Crusade* (London: Colborn, 1847), 124. Disraeli would have been interested that in di Lampedusa's *The Leopard* (1958), the prince of Salina's nephew Tancred says, "If we want things to stay as they are, things will have to change."

3. Charles Darwin, *On the Origin of Species by Means of Natural Selection; or, The Preservation of Favoured Races in the Struggle for Life* (London: John Murray, 1859), 489–90.

4. Francis Darwin, *Charles Darwin: The Foundations of the "Origin of Species," Two Essays Written in 1842 and 1844* (Cambridge: Cambridge University Press, 1909), 254–55.

5. Darwin, *Origin of Species*, 483–84.

6. Ibid., 488–89.

7. James D. Loy and Kent M. Loy, *Emma Darwin: A Victorian Life* (Gainesville: University Press of Florida, 2010).

8. Darwin to Caroline, April 6, 1826, in *The Correspondence of Charles Darwin*, ed. Frederick Burkhardt and Sydney Smith (Cambridge: Cambridge University Press, 1985), 1: 139.

9. Keith Thomson, *The Young Charles Darwin* (New Haven: Yale University Press, 2009), 69–70.

10. *Charles Darwin, Autobiographies*, ed. Michael Neve (London: Penguin, 1905), 49.

11. Henslow, a professor of botany and mineralogy at Cambridge, had the living of the parish church of Cholsey, near Oxford. Passengers on the railway between London and Oxford can see Cholsey Church, a sturdy twelfth-century building built on a Saxon foundation (and, incidentally, the burial place of Agatha Christie), standing lonely in the fields among a few cottages a mile or so to the north of the tracks.

12. *Charles Darwin's Notebooks*, B Notebook, entry 225.

13. The best study of Darwin's illnesses is Ralph Culp, *Darwin's Illness* (Gainesville: University of Florida Press, 2008).

14. *Darwin, Autobiographies*, 55.

15. Emma Wedgwood to Darwin, November 21, 1838, in *Correspondence*, 2:122–23.

16. Darwin, *Autobiographies*, 50.

17. Nora Barlow, ed., *The Autobiography of Charles Darwin* (London: Harcourt, Brace, 1958), 87.

18. Darwin, *Autobiographies*, 50.

19. Ibid., 53–54.

20. John Stevens Henslow, *Descriptive and Physiological Botany* (London: Longmans, 1839), 305.

21. The Reverend Baden Powell devoted a lot of his book *The Order of Nature Considered with Reference to the Claims of Revelation* to demolition of the concept of biblical miracles and denial that miracles occurred in modern times. And he applied this reasoning to the notion of Special Creation. "It is absurd to argue that the introduction of new forms of life, as new species of organized beings, in the successive epochs of the earth's formation, involves a peculiar mysterious power, or supernatural creation, merely because we do not at present know the cause of life, or see new species arise before our eyes, which, it may be added, we never could detect as such if they did." He summarized: "Both the idea of self-existence and that of creation out of nothing are equally and hopelessly beyond the possible grasp of the human faculties, how then can we pretend to reason, or infer anything respecting them? All such fancies must be steadily banished from the domain of real science" (468).

22. *Charles Darwin's Notebooks*, B Notebook, entries 47 and 229.

23. Darwin, *Origin of Species*, 488.

24. Adam Sedgwick to Darwin, November 24, 1859, Darwin Correspondence Project Database, http://www.darwinproject.ac.uk/entry-2548.

25. E. B. Pusey, *Un-science, Not Science, Adverse to Faith: A Sermon Preached before the University on the Twentieth Day after Trinity, 1878* (London: James Parker, 1878), 56. Pusey was too frail to give the sermon himself; it was read for him.

26. Ibid., 10.

27. Ibid., 31.

28. Ibid., 54.

29. Darwin to H. N. Ridley, November 18, 1878, Darwin Correspondence Project Database, http://www.darwinproject.ac.uk/entry-11766.

30. Charles Darwin, *The Descent of Man and Selection in Relation to Sex* (London: Murray, 1871) 1:153.

31. Francis Darwin, *Charles Darwin*, 22.

32. Darwin, *On the Origin of Species*, 185–86.

33. Ibid., 95–96.

34. Ibid., 186.

35. Francis Darwin, *Charles Darwin*, 202.

CHAPTER 6. THE DEVIL AND MR. DARWIN

1. For an up-to-date review of Darwin's writings on religion, see J. David Pleins, *The Evolving God: Charles Darwin on the Naturalness of Religion* (New York: Bloomsbury, 2013).

2. *On the Origin of Species*, 2nd ed. (1860), 480–81. That author probably was Kingsley.

3. Darwin to J. D. Hooker, March 29, 1863, Darwin Correspondence Project Database, http://www.darwinproject.ac.uk/entry-11766.

4. Darwin, Letter to the editor, *Athenaeum*, April 25, 1863, 554–55.

5. Darwin, *Descent of Man*, 613.

6. Darwin to J. D. Hooker, July 17, 1870, Darwin Correspondence Project Database, http://www.darwinproject.ac.uk/entry-7273.

7. Darwin to John Fordyce, May 1, 1879, Darwin Correspondence Project Database. http://www.darwinproject.ac.uk/entry-12041.

8. *Darwin, Autobiographies*, 53.

9. Darwin to Brodie Innes, November 27, 1878, Darwin Correspondence Project Database, http://www.darwinproject.ac.uk/entry-11763.

10. Darwin to E. B. Aveling, October 13, 1880, Darwin Correspondence Project Database, http://www.darwinproject.ac.uk/entry-12757.

11. Darwin, *Descent of Man*, 109.

CHAPTER 7. DEBATES AND ACADEMICS

1. J. D. Hooker, July 2, 1860, Darwin Correspondence Project Database, http://www.darwinproject.ac.uk/entry-2852.

2. Louis Agassiz, *An Essay on Classification* (London: Longmans, Green, 1859), 205.

3. Dorothy Wayman, *Edward Sylvester Morse: A Biography* (Cambridge, Mass.: Harvard University Press, 1942), 120.

4. Asa Gray, "Diagnostic Characters of New Species of Phaenogamous Plants. Collected in Japan by Charles Wright, Botanist of the U.S. North Pacific Exploring Expedition. With Observations upon the Relations of the Japanese Flora to That of North America, and of Other Parts of the Northern Temperate Zone," *Memoirs of the American Academy of Arts and Sciences*, n.s., 6 (1860): 377–452.

5. *Proceedings of the American Academy of Arts and Sciences* 4 (1857–60): 131.

6. Ibid., 133.

7. Ibid., 132

8. Wayman, *Edward Sylvester Morse*, 114.

9. *Proceedings*, 178–79.

10. Ibid., 134.

11. Ibid., 177.

12. Gray to Hooker, January 5, 1860, Darwin Correspondence Project Database, http://www.darwinproject.ac.uk/entry-2638. Gray's letter to Darwin is missing.

13. *Proceedings*, 430 (published out of sequence).

14. Ibid., 413.

15. Ibid., 425–26.

16. Gray to Darwin, April 14, 1871, Darwin Correspondence Project Database, http://www.darwinproject.ac.uk/entry-7683.

CHAPTER 8. CLERICS AND APES

1. Powell, *The Order of Nature*, 468.

2. Keith Thomson, "Huxley, Wilberforce, and the Oxford Museum," *American Scientist* 88 (2000): 210–13.

3. *Athenaeum*, July 1860, 19, www.darwinproject.ac.uk/british-association -meeting-1860.

4. Ibid., 64–65.

5. Ibid.

6. Ibid., 65.

7. Frederick Temple, "A Sermon Preached before the University of Oxford, on Act Sunday, July 1st, 1860," quoted in "Oxford British Association Discussions as Related to Spiritual Questions," *Christian Remembrancer* 40 (1861): 244.

CHAPTER 9. THE DECLINE OF AUTHORITY

1. "Anecdotes of Mary H [Hewson]," *Pennsylvania Magazine of History and Biography* 19 (1895): 407–9. The anecdote is also recorded by Benjamin Rush in his *Commonplace Book*.

2. http://www.pewforum.org/2008/06/01/u-s-religious-landscape-survey -religious-beliefs-and-practices/.

3. http://www.pewforum.org/2010/09/28/u-s-religious-knowledge-survey/; www.gallup.com/poll/21814/Evolution-Creationism-Intelligent-Design.aspx; http://www.bloomberg.com/apps/news?pid=newsarchive&sid=a8lZO7W8yrLY; https://www.barna.org/barna-update/faith-spirituality/260-most-american -christians-do-not-believe-that-satan-or-the-holy-spirit-exis#.Uvep9_s1CS0 (2009).

4. http://www.bloomberg.com/apps/news?pid=newsarchive&sid=a8lZO7 W8yrLY; http://www.pewforum.org/2008/06/01/u-s-religious-landscape-survey -religious-beliefs-and-practices/.

5. *New York Times*, December 8, 2010.

6. C. Kirk Hardaway, Penny Long Marler, and Mark Chaves, "What the Polls Don't Show: A Closer Look at U.S. Church Attendance," *American Sociological Review* 58, no. 6 (1993): 741–52.

7. George F. Bishop, Randall K. Thomas, and Jason A. Wood, "Americans' Scientific Knowledge and Beliefs about Human Evolution in the Year of Darwin," *National Centre for Science Education Reports* 30 (2010): 16–18, http:// www.pewforum.org/2013/12/30/publics-views-on-human-evolution/.

8. Hardaway, Marler, and Chaves, "What the Polls Don't Show."

9. Pew Forum on Religion and Public Life, http://religions.pewforum.org /reports; Association of Religion Data Archives, http://www.thearda.com/quick stats/qs_107.asp (2008).

10. *Austin (Tex.) Chronicle*, June 6, 2012.

CHAPTER 10. A WAY FORWARD?

1. See, for example, Don K. Price, *The Scientific Estate* (Cambridge, Mass.: Harvard University Press, 1965).

2. John Hedley Brooke, *Science and Religion: Some Historical Perspectives* (Cambridge: Cambridge University Press, 1991).

3. Keith Thomas, *Religion and the Decline of Magic* (London: Weidenfeld and Nicholson, 1971), 206.

4. It also has to be noted that neither science nor religion have entirely removed magic from our cultures; they have merely provided an alter ego for it in,

for example, sainthood from miracles, holy water, sacred relics, and sites with healing powers.

5. Samuel H. Thomson, *Geology as an Interpreter of Scripture: An Address Delivered at the Anniversary of the Union Literary and Philalethean Societies of Hanover College*, August 5, 1858 (Cincinnati: Franklin Steam, 1858), 32.

6. Stephen J. Gould, *Rocks of Ages: Science and Religion in the Fullness of Life* (New York: Ballantine, 1999).

7. It is essential to distinguish between contraception, which is the prevention of the creation of a fertilized egg, and abortion, which even extremely early on means the termination of pregnancy. The condom and morning-after pill are contraceptives and, however one defines the beginning of life, do not go against the injunction "Thou shalt not kill." Religious zealots find it convenient to blur this distinction.

8. Save the Children, *Every Woman's Right: How Family Planning Saves Lives* (London, 2012).

9. Catholic News Agency, "Pope Benedict XVI: Full Text concerning HIV/ AIDS," March 19, 2009, http://www.ewtn.com/vnews/getstory.asp?number= 94418.

10. *Philadelphia Inquirer*, October 7, 2012.

11. *New York Times*, March 14, 2012.

APPENDIX

1. Wilberforce teased his audience by assuming that they would not need a translation for this passage from Lucretius's *De Rerum Natura*. The passage in question is preceded by reference to the fact that the ability of species to propagate their kind is dependent on access to food and that both male and female must have the appropriate organs for "the exchange of earthly delights," as the early eighteenth-century translator Thomas Creech delicately put it ("Pabula primum ut sint, genitalia deinde per artus semina qua possint membris manare remissis, feminaque ut maribus coniungi possit, habere, mutua qui mutent inter se gaudia uterque"). Perhaps that was a bit racy for Wilberforce, who began his quotation from the poem at the next lines, which can be translated roughly to the effect that "meanwhile, over time, many species have failed to propagate themselves. Those that survive today to bear young have been preserved through cunning, strength, or speed. Many also are protected by man for their utility."

Index

View of the Evidences of Christianity, A
  (Paley), 74
volcanoes and earthquakes, 46, 52, 76

Wallace, Alfred Russell, 108
watch metaphor, 17
Wedgwood, Emma. *See* Darwin,
  Emma Wedgwood
Whewell, William, 96
White, Gilbert, 190n.10
Whitehurst, John, 34–35
Wilberforce, Samuel, 107–108, 126–
  127, 128–134
Wilberforce's Oxford address, 169–

188; admiration for *On the Origin of
Species*, 170–171; on artificial selection and variation, 177–180; on conclusion that all life descended from
one primordial form, 172–174; on
Darwin's use of time to accomplish
transmutation, 182–185; on evidence
for favorable variation, 180–182; on
humble-bees and red clover, 171–172;
on natural selection, 174–176; religious objections to Darwin's theory,
185–188
Williams, Rowland, 141–142
Wordsworth, William, ix–xi